Introduction to Electrodynamics

Introduction to Electrodynamics

Edited by
Roger Carroll

Larsen & Keller
www.larsen-keller.com

Introduction to Electrodynamics
Edited by Roger Carroll
ISBN: 978-1-63549-706-9 (Hardback)

Larsen & Keller

Published by Larsen and Keller Education,
5 Penn Plaza,
19th Floor,
New York, NY 10001, USA

Cataloging-in-Publication Data

Introduction to electrodynamics / edited by Roger Carroll.
p. cm.
Includes bibliographical references and index.
ISBN 978-1-63549-706-9
1. Electrodynamics. 2. Dynamics. 3. Physics. I. Carroll, Roger.
QC631 .I58 2018
537.6--dc23

For more information regarding Larsen and Keller Education and its products, please visit the publisher's website www.larsen-keller.com

Table of Contents

Preface

The study of the relationship between electric charge and currents by using Newton's model is referred to as electrodynamics. It is a sub-discipline of theoretical physics. The main concepts covered under this subject are electric field, Lorentz field, general field equations, electromagnetic waves, etc. This book is compiled in such a manner, that it will provide in-depth knowledge about the theory and practice of electrodynamics. Some of the diverse topics covered in this book address the varied branches that fall under this category. For all those who are interested in electrodynamics, this textbook can prove to be an essential guide.

A foreword of all chapters of the book is provided below:

Chapter 1 - Classical electrodynamics studies electric current and charge using the Newtonian model. Electromagnetic radiation comprises radio waves, infrared, ultraviolet, light, and gamma radiation. Waves carrying energy are called electromagnetic waves. The chapter explores the idea of Maxwell's equations, which forms the basis of classical electromagnetism. This set of partial differential equations explain the generation of magnetic and electric fields. The section on electromagnetic waves offers an insightful focus, keeping in mind the complex subject matter; **Chapter 2** - The physical field that is produced by electrically charged objects is known as electromagnetic field. This section has been carefully written to provide an easy understanding of the varied facets of electrodynamics. This chapter explores the notions such as stationary states and plane waves and other concepts that form the basics of electrodynamics. The chapter strategically encompasses and incorporates the major components and key concepts of electromagnetic fields and waves, providing a complete understanding; **Chapter 3** - Lorentz force can be understood as the force on an electric charge q moving with velocity v in a magnetic field B and electric field E. Lorentz force, when taken with Maxwell's equation, establishes the underlying principles of electric, optical and radio technologies. This chapter discusses in detail the theories and methodologies related to relativity and electrodynamics; **Chapter 4** - Special relativity establishes the relation between space and time. It forms the precise model of motion at any speed. Special relativity provides formulas of the change that occurs in electromagnetic objects under Lorentz transformation as well the relationship between electricity and magnetism. The aim of this section is to explore the special relativity and electromagnetism. These topics are crucial for a complete understanding of the subject.

At the end, I would like to thank all the people associated with this book devoting their precious time and providing their valuable contributions to this book. I would also like to express my gratitude to my fellow colleagues who encouraged me throughout the process.

Editor

1

Electromagnetic Waves

Classical electrodynamics studies electric current and charge using the Newtonian model. Electromagnetic radiation comprises radio waves, infrared, ultraviolet, light, and gamma radiation. Waves carrying energy are called electromagnetic waves. The chapter explores the idea of Maxwell's equations, which forms the basis of classical electromagnetism. This set of partial differential equations explain the generation of magnetic and electric fields. The section on electromagnetic waves offers an insightful focus, keeping in mind the complex subject matter.

Classical Electromagnetism

Classical electromagnetism or classical electrodynamics is a branch of theoretical physics that studies the interactions between electric charges and currents using an extension of the classical Newtonian model. The theory provides an excellent description of electromagnetic phenomena whenever the relevant length scales and field strengths are large enough that quantum mechanical effects are negligible. For small distances and low field strengths, such interactions are better described by quantum electrodynamics.

Fundamental physical aspects of classical electrodynamics are presented in many texts, such as those by Feynman, Leighton and Sands, Griffiths, Panofsky and Phillips, and Jackson.

History

The physical phenomena that electromagnetism describes have been studied as separate fields since antiquity. For example, there were many advances in the field of optics centuries before light was understood to be an electromagnetic wave. However, the theory of electromagnetism, as it is currently understood, grew out of Michael Faraday's experiments suggesting an electromagnetic field and James Clerk Maxwell's use of differential equations to describe it in his *A Treatise on Electricity and Magnetism* (1873).

Lorentz force

The electromagnetic field exerts the following force (often called the Lorentz force) on charged particles:

$$F = qE + qv \times B$$

where all boldfaced quantities are vectors: F is the force that a particle with charge *q* experiences, E is the electric field at the location of the particle, **v** is the velocity of the particle, B is the magnetic field at the location of the particle.

The above equation illustrates that the Lorentz force is the sum of two vectors. One is the cross product of the velocity and magnetic field vectors. Based on the properties of the cross product, this produces a vector that is perpendicular to both the velocity and magnetic field vectors. The other vector is in the same direction as the electric field. The sum of these two vectors is the Lorentz force.

Therefore, in the absence of a magnetic field, the force is in the direction of the electric field, and the magnitude of the force is dependent on the value of the charge and the intensity of the electric field. In the absence of an electric field, the force is perpendicular to the velocity of the particle and the direction of the magnetic field. If both electric and magnetic fields are present, the Lorentz force is the sum of both of these vectors.

The Electric Field E

The electric field E is defined such that, on a stationary charge:

$$F = q_0 E$$

where q_o is what is known as a test charge. The size of the charge doesn't really matter, as long as it is small enough not to influence the electric field by its mere presence. What is plain from this definition, though, is that the unit of E is N/C (newtons per coulomb). This unit is equal to V/m (volts per meter).

In electrostatics, where charges are not moving, around a distribution of point charges, the forces determined from Coulomb's law may be summed. The result after dividing by q_o is:

$$E(r) = \frac{1}{4pe_0} \sum_{i=1}^{n} \frac{q_i (r - r_i)}{|r - r_i|^3}$$

where *n* is the number of charges, q_i is the amount of charge associated with the *i*th charge, r_i is the position of the *i*th charge, r is the position where the electric field is being determined, and ε_o is the electric constant.

If the field is instead produced by a continuous distribution of charge, the summation becomes an integral:

$$E(r) = \frac{1}{4pe_0} \int \frac{r(r')(r - r')}{|r - r'|^3} d^3r'$$

where $\rho(\mathrm{r}')$ is the charge density and $\mathrm{r}-\mathrm{r}'$ is the vector that points from the volume element $\mathrm{d}^3\mathrm{r}'$ to the point in space where E is being determined.

Both of the above equations are cumbersome, especially if one wants to determine E as a function of position. A scalar function called the electric potential can help. Electric potential, also called voltage (the units for which are the volt), is defined by the line integral

$$\varphi(r) = -\int_C \mathrm{E} \times \mathrm{dl}$$

where *φ(r)* is the electric potential, and *C* is the path over which the integral is being taken.

Unfortunately, this definition has a caveat. From Maxwell's equations, it is clear that $\nabla \times \mathrm{E}$ is not always zero, and hence the scalar potential alone is insufficient to define the electric field exactly. As a result, one must add a correction factor, which is generally done by subtracting the time derivative of the A vector potential described below. Whenever the charges are quasistatic, however, this condition will be essentially met.

From the definition of charge, one can easily show that the electric potential of a point charge as a function of position is:

$$\varphi(\mathrm{r}) = \frac{1}{4\pi\varepsilon_0} \sum_{i=1}^{n} \frac{q_i}{|\mathrm{r}-\mathrm{r}_i|}$$

where q is the point charge's charge, r is the position at which the potential is being determined, and r_i is the position of each point charge. The potential for a continuous distribution of charge is:

$$\varphi(\mathrm{r}) = \frac{1}{4\pi\varepsilon_0} \int \frac{\rho(\mathrm{r}')}{|\mathrm{r}-\mathrm{r}'|} \mathrm{d}^3\mathrm{r}'$$

where $\rho(\mathrm{r}')$ is the charge density, and $\mathrm{r}-\mathrm{r}'$ is the distance from the volume element $\mathrm{d}^3\mathrm{r}'$ to point in space where φ is being determined.

The scalar φ will add to other potentials as a scalar. This makes it relatively easy to break complex problems down in to simple parts and add their potentials. Taking the definition of φ backwards, we see that the electric field is just the negative gradient (the del operator) of the potential.

Or: $\mathrm{E}(\mathrm{r}) = -\nabla\varphi(\mathrm{r})$.

From this formula it is clear that E can be expressed in V/m (volts per meter).

Electromagnetic Waves

A changing electromagnetic field propagates away from its origin in the form of a wave. These waves travel in vacuum at the speed of light and exist in a wide spectrum of wavelengths. Examples of the dynamic fields of electromagnetic radiation (in order of increasing frequency): radio waves, microwaves, light (infrared, visible light and ultraviolet), x-rays and gamma rays. In the field of particle physics this electromagnetic radiation is the manifestation of the electromagnetic interaction between charged particles.

General Field Equations

As simple and satisfying as Coulomb's equation may be, it is not entirely correct in the context of classical electromagnetism. Problems arise because changes in charge distributions require a non-zero amount of time to be "felt" elsewhere (required by special relativity).

For the fields of general charge distributions, the retarded potentials can be computed and differentiated accordingly to yield Jefimenko's Equations.

Retarded potentials can also be derived for point charges, and the equations are known as the Liénard–Wiechert potentials. The scalar potential is:

$$\varphi = \frac{1}{4\pi\varepsilon_0} \frac{q}{\left|\mathrm{r}-\mathrm{r}_q(t_{ret})\right| - \frac{\mathrm{v}_q(t_{ret})}{c}\cdot(\mathrm{r}-\mathrm{r}_q(t_{ret}))}$$

where q is the point charge's charge and r is the position. r_q and v_q are the position and velocity of the charge, respectively, as a function of retarded time. The vector potential is similar:

$$\mathrm{A} = \frac{\mu_0}{4\pi} \frac{q\,\mathrm{v}_q(t_{ret})}{\left|\mathrm{r}-\mathrm{r}_q(t_{ret})\right| - \frac{\mathrm{v}_q(t_{ret})}{c}\cdot(\mathrm{r}-\mathrm{r}_q(t_{ret}))}.$$

These can then be differentiated accordingly to obtain the complete field equations for a moving point particle.

Models

Branches of classical electromagnetism such as optics, electrical and electronic engineering consist of a collection of relevant mathematical models of different degrees of simplification and idealization to enhance the understanding of specific electrodynamics phenomena, cf. An electrodynamics phenomenon is determined by the particular fields, specific densities of electric charges and currents, and the particular transmission medium. Since there are infinitely many of them, in modeling there is a need for some typical, representative:

(a) electrical charges and currents, e.g. moving pointlike charges and electric and magnetic dipoles, electric currents in a conductor etc.;

(b) electromagnetic fields, e.g. voltages, the Liénard–Wiechert potentials, the monochromatic plane waves, optical rays; radio waves, microwaves, infrared radiation, visible light, ultraviolet radiation, X-rays, gamma rays etc.;

(c) transmission media, e.g. electronic components, antennas, electromagnetic waveguides, flat mirrors, mirrors with curved surfaces convex lenses, concave lenses; resistors, inductors, capacitors, switches; wires, electric and optical cables, transmission lines, integrated circuits etc.;

all of which have only few variable characteristics.

Electromagnetic Radiation

In physics, electromagnetic radiation (EM radiation or EMR) refers to the waves (or their quanta, photons) of the electromagnetic field, propagating (radiating) through space carrying electromagnetic radiant energy. It includes radio waves, microwaves, infrared, (visible) light, ultraviolet, X-, and gamma radiation.

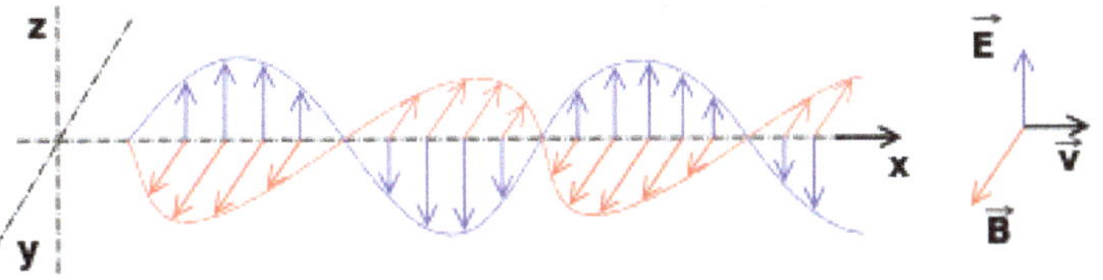

The electromagnetic waves that compose electromagnetic radiation can be imagined as a self-propagating transverse oscillating wave of electric and magnetic fields. This diagram shows a plane linearly polarized EMR wave propagating from left to right (X axis). The electric field is in a vertical plane (Z axis) and the magnetic field in a horizontal plane (Y axis). The electric and magnetic fields in EMR waves are always in phase and at 90 degrees to each other.

Classically, electromagnetic radiation consists of electromagnetic waves, which are synchronized oscillations of electric and magnetic fields that propagate at the speed of light through a vacuum. The oscillations of the two fields are perpendicular to each other and perpendicular to the direction of energy and wave propagation, forming a transverse wave. The wavefront of electromagnetic waves emitted from a point source (such as a lightbulb) is a sphere. The position of an electromagnetic wave within the electromagnetic spectrum can be characterized by either its frequency of oscillation or its wavelength. The electromagnetic spectrum includes, in order of increasing frequency and decreasing wavelength: radio waves, microwaves, infrared radiation, visible light, ultraviolet radiation, X-rays and gamma rays.

Electromagnetic waves are produced whenever charged particles are accelerated, and these waves can subsequently interact with other charged particles. EM waves carry energy, momentum and angular momentum away from their source particle and can impart those quantities to matter with which they interact. Quanta of EM waves are called photons, whose rest mass is zero, but whose energy, or equivalent total (relativistic) mass, is not zero so they are still affected by gravity. Electromagnetic radiation is associated with those EM waves that are free to propagate themselves ("radiate") without the continuing influence of the moving charges that produced them, because they have achieved sufficient distance from those charges. Thus, EMR is sometimes referred to as the far field. In this language, the *near field* refers to EM fields near the charges and current that directly produced them, specifically, electromagnetic induction and electrostatic induction phenomena.

In the quantum theory of electromagnetism, EMR consists of photons, the elementary particles responsible for all electromagnetic interactions. Quantum effects provide additional sources of EMR, such as the transition of electrons to lower energy levels in an atom and black-body radiation. The energy of an individual photon is quantized and is greater for photons of higher frequency. This relationship is given by Planck's equation $E = hv$, where E is the energy per photon, v is the frequency of the photon, and h is Planck's constant. A single gamma ray photon, for example, might carry ~100,000 times the energy of a single photon of visible light.

The effects of EMR upon chemical compounds and biological organisms depend both upon the radiation's power and its frequency. EMR of visible or lower frequencies (i.e., visible light, infrared, microwaves, and radio waves) is called *non-ionizing radiation*, because its photons do not individually have enough energy to ionize atoms or molecules. The effects of these radiations on chemical systems and living tissue are caused primarily by heating effects from the combined energy transfer of many photons. In contrast, high ultraviolet, X-rays and gamma rays are called *ionizing radiation* since individual photons of such high frequency have enough energy to ionize molecules or break chemical bonds. These radiations have the ability to cause chemical reactions and damage living cells beyond that resulting from simple heating, and can be a health hazard.

Physics

Maxwell's Equations

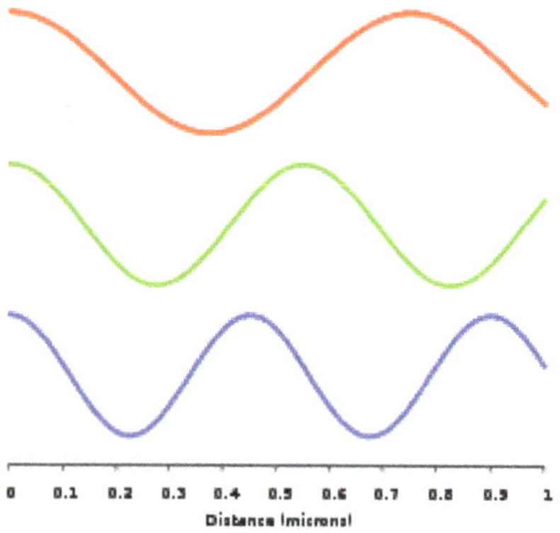

Shows the relative wavelengths of the electromagnetic waves of three different colours of light (blue, green, and red) with a distance scale in micrometers along the x-axis.

Maxwell derived a wave form of the electric and magnetic equations, thus uncovering the wave-like nature of electric and magnetic fields and their symmetry. Because the speed of EM waves predicted by the wave equation coincided with the measured speed of light, Maxwell concluded that light itself is an EM wave. Maxwell's equations were confirmed by Heinrich Hertz through experiments with radio waves.

According to Maxwell's equations, a spatially varying electric field is always associated with a magnetic field that changes over time. Likewise, a spatially varying magnetic field is associated with specific changes over time in the electric field. In an electromagnetic wave, the changes in the electric field are always accompanied by a wave in the magnetic field in one direction, and vice versa. This relationship between the two occurs without either type field causing the other; rather, they occur together in the same way that time and space changes occur together and are interlinked in special relativity. In fact, magnetic fields may be viewed as relativistic distortions of electric fields, so the close relationship between space and time changes here is more than an analogy. Together, these fields form a propagating electromagnetic wave, which moves out into space and need never again affect the source. The distant EM field formed in this way by the acceleration of a charge carries energy with it that "radiates" away through space, hence the term.

Near and Far Fields

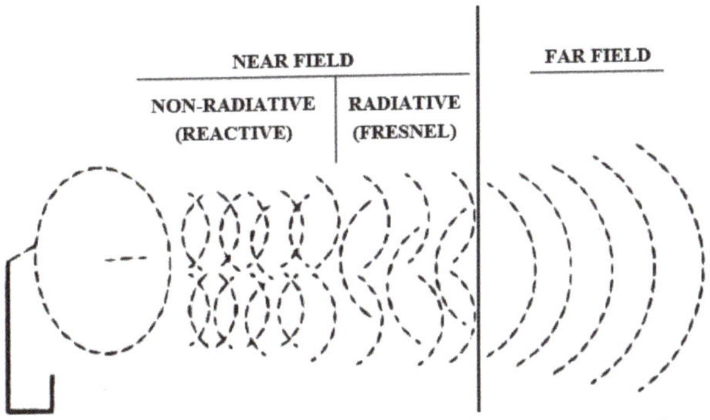

In electromagnetic radiation (such as microwaves from an antenna, shown here) the term applies only to the parts of the electromagnetic field that radiate into infinite space and decrease in intensity by an inverse-square law of power, so that the total radiation energy that crosses through an imaginary spherical surface is the same, no matter how far away from the antenna the spherical surface is drawn. Electromagnetic radiation includes the far field part of the electromagnetic field around a transmitter. A part of the "near-field" close to the transmitter, forms part of the changing electromagnetic field, but does not count as electromagnetic radiation.

Maxwell's equations established that some charges and currents ("sources") produce a local type of electromagnetic field near them that does *not* have the behaviour of EMR. Currents directly produce a magnetic field, but it is of a magnetic dipole type that dies out with distance from the current. In a similar manner, moving charges pushed apart in a conductor by a changing electrical potential (such as in an antenna) produce an electric dipole type electrical field, but this also declines with distance. These fields make up the near-field near the EMR source. Neither of these behaviours are responsible for EM radiation. Instead, they cause electromagnetic field behaviour that only efficiently transfers power to a receiver very close to the source, such as the magnetic induction inside a transformer, or the feedback behaviour that happens close to the coil of a metal detector. Typically, near-fields have a powerful effect on their own sources, causing an increased "load" (decreased electrical reactance) in the source or transmitter, whenever energy is withdrawn from the EM field by a receiver. Otherwise, these fields do not "propagate" freely out into space, carrying their energy away without distance-limit, but rather oscillate, returning their energy to the transmitter if it is not received by a receiver.

By contrast, the EM far-field is composed of *radiation* that is free of the transmitter in the sense that (unlike the case in an electrical transformer) the transmitter requires the same power to send these changes in the fields out, whether the signal is immediately picked up or not. This distant part of the electromagnetic field *is* "electromagnetic radiation" (also called the far-field). The far-fields propagate (radiate) without allowing the transmitter to affect them. This causes them to be independent in the sense that their existence and their energy, after they have left the transmitter, is completely independent of both transmitter and receiver. Because such waves conserve the amount of energy they transmit through any spherical boundary surface drawn around their source, and because such surfaces have an area that is defined by the square of the distance from the source, the power of EM radiation always varies according to an inverse-square law. This is in contrast to dipole parts of the EM field close to the source (the near-field), which varies in power according to an inverse cube power law, and thus does *not* transport a conserved amount of energy over distances, but instead fades with distance, with its energy (as noted) rapidly returning to the transmitter or absorbed by a nearby receiver (such as a transformer secondary coil).

The far-field (EMR) depends on a different mechanism for its production than the near-field, and upon different terms in Maxwell's equations. Whereas the magnetic part of the near-field is due to currents in the source, the magnetic field in EMR is due only to the local change in the electric field. In a similar way, while the electric field in the near-field is due directly to the charges and charge-separation in the source, the electric field in EMR is due to a change in the local magnetic field. Both processes for producing electric and magnetic EMR fields have a different dependence on distance than do near-field dipole electric and magnetic fields. That is why the EMR type of EM field becomes dominant in power "far" from sources. The term "far from sources" refers to how far from the source (moving at the speed of light) any portion of the outward-moving EM field is located, by the time that source currents are changed by the varying source potential, and the source has therefore begun to generate an outwardly moving EM field of a different phase.

A more compact view of EMR is that the far-field that composes EMR is generally that part of the EM field that has traveled sufficient distance from the source, that it has become completely disconnected from any feedback to the charges and currents that were originally responsible for it. Now independent of the source charges, the EM field, as it moves farther away, is dependent only upon the accelerations of the charges that produced it. It no longer has a strong connection to the direct fields of the charges, or to the velocity of the charges (currents).

In the Liénard–Wiechert potential formulation of the electric and magnetic fields due to motion of a single particle (according to Maxwell's equations), the terms associated with acceleration of the particle are those that are responsible for the part of the field that is regarded as electromagnetic radiation. By contrast, the term associated with the changing static electric field of the particle and the magnetic term that results from the particle's uniform velocity, are both associated with the electromagnetic near-field, and do not comprise EM radiation.

Properties

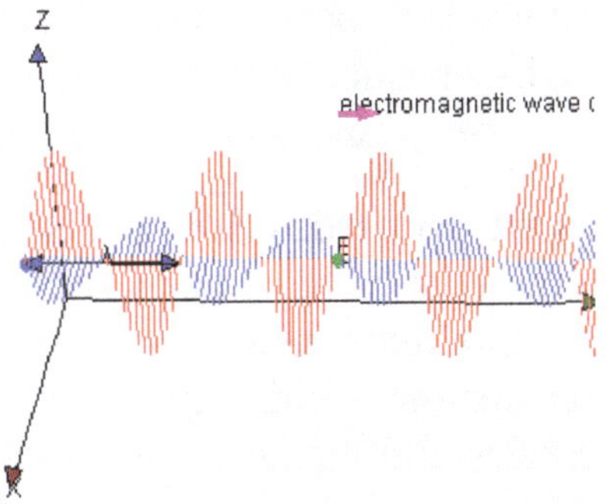

Electromagnetic waves can be imagined as a self-propagating transverse oscillating wave of electric and magnetic fields. Note that the electric and magnetic fields in such a wave are in-phase with each other, reaching minima and maxima together

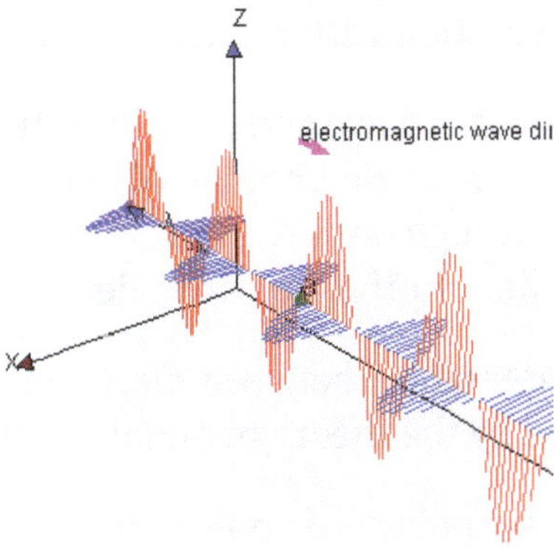

An alternate view of the wave shown above.

Electrodynamics is the physics of electromagnetic radiation, and electromagnetism is the physical phenomenon associated with the theory of electrodynamics. Electric and magnetic fields obey the properties of superposition. Thus, a field due to any particular particle or time-varying electric or magnetic field contributes to the fields present in the same space due to other causes. Further, as they are vector fields, all magnetic and electric field vectors add together according to vector addition. For example, in optics two or more coherent lightwaves may interact and by constructive or destructive interference yield a resultant irradiance deviating from the sum of the component irradiances of the individual lightwaves.

Since light is an oscillation it is not affected by traveling through static electric or magnetic fields in a linear medium such as a vacuum. However, in nonlinear media, such as some crystals, interactions can occur between light and static electric and magnetic fields — these interactions include the Faraday effect and the Kerr effect.

In refraction, a wave crossing from one medium to another of different density alters its speed and direction upon entering the new medium. The ratio of the refractive indices of the media determines the degree of refraction, and is summarized by Snell's law. Light of composite wavelengths (natural sunlight) disperses into a visible spectrum passing through a prism, because of the wavelength-dependent refractive index of the prism material (dispersion); that is, each component wave within the composite light is bent a different amount.

EM radiation exhibits both wave properties and particle properties at the same time. Both wave and particle characteristics have been confirmed in many experiments. Wave characteristics are more apparent when EM radiation is measured over relatively large timescales and over large distances while particle characteristics are more evident when measuring small timescales and distances. For example, when electromagnetic radiation is absorbed by matter, particle-like properties will be more obvious when the average number of photons in the cube of the relevant wavelength is much smaller than 1. It is not too difficult to experimentally observe non-uniform deposition of energy when light is absorbed, however this alone is not evidence of "particulate" behavior. Rather, it reflects the quantum nature of *matter*. Demonstrating that the light itself is quantized, not merely its interaction with matter, is a more subtle affair.

Some experiments display both the wave and particle natures of electromagnetic waves, such as the self-interference of a single photon. When a single photon is sent through an interferometer, it passes through both paths, interfering with itself, as waves do, yet is detected by a photomultiplier or other sensitive detector only once.

A quantum theory of the interaction between electromagnetic radiation and matter such as electrons is described by the theory of quantum electrodynamics.

Electromagnetic waves can be polarized, reflected, refracted, diffracted or interfere with each other.

Wave Model

Electromagnetic radiation is a transverse wave, meaning that its oscillations are perpendicular to the direction of energy transfer and travel. The electric and magnetic parts of the field stand in a fixed ratio of strengths in order to satisfy the two Maxwell equations that specify how one is produced from the other. These E and B fields are also in phase, with both reaching maxima and minima at the same points in space. A common misconception is that the E and B fields in electromagnetic radiation are out of phase because a change in one produces the other, and this would produce a phase difference between them as sinusoidal functions (as indeed happens in electromagnetic induction, and in the near-field close to antennas). However, in the far-field EM radiation which is described by the two source-free Maxwell curl operator equations, a more correct description is that a time-change in one type of field is proportional to a space-change in the other. These derivatives require that the E and B fields in EMR are in-phase.

An important aspect of light's nature is its frequency. The frequency of a wave is its rate of oscillation and is measured in hertz, the SI unit of frequency, where one hertz is equal to one oscillation per second. Light usually has multiple frequencies that sum to form the resultant wave. Different frequencies undergo different angles of refraction, a phenomenon known as dispersion.

A wave consists of successive troughs and crests, and the distance between two adjacent crests or troughs is called the wavelength. Waves of the electromagnetic spectrum vary in size, from very long radio waves the size of buildings to very short gamma rays smaller than atom nuclei. Frequency is inversely proportional to wavelength, according to the equation:

$$v = f\lambda$$

where v is the speed of the wave (c in a vacuum, or less in other media), f is the frequency and λ is the wavelength. As waves cross boundaries between different media, their speeds change but their frequencies remain constant.

Electromagnetic waves in free space must be solutions of Maxwell's electromagnetic wave equation. Two main classes of solutions are known, namely plane waves and spherical waves. The plane waves may be viewed as the limiting case of spherical waves at a very large (ideally infinite) distance from the source. Both types of waves can have a waveform which is an arbitrary time function (so long as it is sufficiently differentiable to conform to the wave equation). As with any time function, this can be decomposed by means of Fourier analysis into its frequency spectrum, or individual sinusoidal components, each of which contains a single frequency, amplitude and phase. Such a component wave is said to be *monochromatic*. A monochromatic electromagnetic wave can be characterized by its frequency or wavelength, its peak amplitude, its phase relative to some reference phase, its direction of propagation and its polarization.

Interference is the superposition of two or more waves resulting in a new wave pattern. If the fields have components in the same direction, they constructively interfere, while opposite directions cause destructive interference. An example of interference caused by EMR is electromagnetic interference (EMI) or as it is more commonly known as, radio-frequency interference (RFI). Additionally, multiple polarization signals can be combined (i.e. interfered) to form new states of polarization, which is known as parallel polarization state generation.

The energy in electromagnetic waves is sometimes called radiant energy.

Particle Model and Quantum Theory

An anomaly arose in the late 19th century involving a contradiction between the wave theory of light and measurements of the electromagnetic spectra that were being emitted by thermal radiators known as black bodies. Physicists struggled with this problem, which later became known as the ultraviolet catastrophe, unsuccessfully for many years. In 1900, Max Planck developed a new theory of black-body radiation that explained the observed spectrum. Planck's theory was based on the idea that black bodies emit light (and other electromagnetic radiation) only as discrete bundles or packets of energy. These packets were called quanta. Later, Albert Einstein proposed that light quanta be regarded as real particles. Later the particle of light was given the name photon, to correspond with other particles being described around this time, such as the electron and proton. A photon has an energy, E, proportional to its frequency, f, by

$$E = hf = \frac{hc}{\lambda}$$

where h is Planck's constant, λ is the wavelength and c is the speed of light. This is sometimes known as the Planck–Einstein equation. In quantum theory the energy of the photons is thus directly proportional to the frequency of the EMR wave.

Likewise, the momentum p of a photon is also proportional to its frequency and inversely proportional to its wavelength:

$$p = \frac{E}{c} = \frac{hf}{c} = \frac{h}{\lambda}.$$

The source of Einstein's proposal that light was composed of particles (or could act as particles in some circumstances) was an experimental anomaly not explained by the wave theory: the photoelectric effect, in which light striking a metal surface ejected electrons from the surface, causing an electric current to flow across an applied voltage. Experimental measurements demonstrated that the energy of individual ejected electrons was proportional to the *frequency*, rather than the *intensity*, of the light. Furthermore, below a certain minimum frequency, which depended on the particular metal, no current would flow regardless of the intensity. These observations appeared to contradict the wave theory, and for years physicists tried in vain to find an explanation. In 1905, Einstein explained this puzzle by resurrecting the particle theory of light to explain the observed effect. Because of the preponderance of evidence in favor of the wave theory, however, Einstein's ideas were met initially with great skepticism among established physicists. Eventually Einstein's explanation was accepted as new particle-like behavior of light was observed, such as the Compton effect.

As a photon is absorbed by an atom, it excites the atom, elevating an electron to a higher energy level (one that is on average farther from the nucleus). When an electron in an excited molecule or atom descends to a lower energy level, it emits a photon of light at a frequency corresponding to the energy difference. Since the energy levels of electrons in atoms are discrete, each element and each molecule emits and absorbs its own characteristic frequencies. Immediate photon emission is called fluorescence, a type of photoluminescence. An example is visible light emitted from fluorescent paints, in response to ultraviolet (blacklight). Many other fluorescent emissions are known in spectral bands other than visible light. Delayed emission is called phosphorescence.

Wave–particle Duality

The modern theory that explains the nature of light includes the notion of wave–particle duality. More generally, the theory states that everything has both a particle nature and a wave nature, and various experiments can be done to bring out one or the other. The particle nature is more easily discerned using an object with a large mass. A bold proposition by Louis de Broglie in 1924 led the scientific community to realize that electrons also exhibited wave–particle duality.

Wave and Particle Effects of Electromagnetic Radiation

Together, wave and particle effects fully explain the emission and absorption spectra of EM radiation. The matter-composition of the medium through which the light travels determines the nature of the absorption and emission spectrum. These bands correspond to the allowed energy levels in the atoms. Dark bands in the absorption spectrum are due to the atoms in an intervening medium between source and observer. The atoms absorb certain frequencies of the light between emitter and detector/eye, then emit them in all directions. A dark band appears to the detector, due to the radiation scattered out of the beam. For instance, dark bands in the light emitted by a distant star are due to the atoms in the star's atmosphere. A similar phenomenon occurs for emission, which is seen when an emitting gas glows due to excitation of the atoms from any mechanism, including heat. As electrons descend to lower energy levels, a spectrum is emitted that represents the jumps between the energy levels of the electrons, but lines are seen because again emission happens only at particular energies after excitation. An example is the emission spectrum of nebulae. Rapidly moving electrons are most sharply accelerated when they encounter a region of force, so they are responsible for producing much of the highest frequency electromagnetic radiation observed in nature.

These phenomena can aid various chemical determinations for the composition of gases lit from behind (absorption spectra) and for glowing gases (emission spectra). Spectroscopy (for example) determines what chemical elements comprise a particular star. Spectroscopy is also used in the determination of the distance of a star, using the red shift.

Propagation Speed

Any electric charge that accelerates, or any changing magnetic field, produces electromagnetic radiation. Electromagnetic information about the charge travels at the speed of light. Accurate treatment thus incorporates a concept known as retarded time, which adds to the expressions for the electrodynamic electric field and magnetic field. These extra terms are responsible for electromagnetic radiation.

When any wire (or other conducting object such as an antenna) conducts alternating current, electromagnetic radiation is propagated at the same frequency as the current. In many such situations it is possible to identify an electrical dipole moment that arises from separation of charges due to the exciting electrical potential, and this dipole moment oscillates in time, as the charges move back and forth. This oscillation at a given frequency gives rise to changing electric and magnetic fields, which then set the electromagnetic radiation in motion.

At the quantum level, electromagnetic radiation is produced when the wavepacket of a charged particle oscillates or otherwise accelerates. Charged particles in a stationary state do not move, but a superposition of such states may result in a transition state that has an electric dipole moment that oscillates in time. This oscillating dipole moment is responsible for the phenomenon of radiative transition between quantum states of a charged particle. Such states occur (for example) in atoms when photons are radiated as the atom shifts from one stationary state to another.

As a wave, light is characterized by a velocity (the speed of light), wavelength, and frequency. As particles, light is a stream of photons. Each has an energy related to the frequency of the wave given by Planck's relation $E = hf$, where E is the energy of the photon, $h = 6.626 \times 10^{-34}$ J·s is Planck's constant, and f is the frequency of the wave.

One rule is obeyed regardless of circumstances: EM radiation in a vacuum travels at the speed of light, *relative to the observer*, regardless of the observer's velocity. (This observation led to Einstein's development of the theory of special relativity.)

In a medium (other than vacuum), velocity factor or refractive index are considered, depending on frequency and application. Both of these are ratios of the speed in a medium to speed in a vacuum.

Special Theory of Relativity

By the late nineteenth century, various experimental anomalies could not be explained by the simple wave theory. One of these anomalies involved a controversy over the speed of light. The speed of light and other EMR predicted by Maxwell's equations did not appear unless the equations were modified in a way first suggested by FitzGerald and Lorentz, or else otherwise that speed would depend on the speed of observer relative to the "medium" (called luminiferous aether) which supposedly "carried" the electromagnetic wave (in a manner analogous to the way air carries sound waves). Experiments failed to find any observer effect. In 1905, Einstein proposed that space and time appeared to be velocity-changeable entities for light propagation and all other processes and laws. These changes accounted for the constancy of the speed of light and all electromagnetic radiation, from the viewpoints of all observers—even those in relative motion.

History of Discovery

Electromagnetic radiation of wavelengths other than those of visible light were discovered in the early 19th century. The discovery of infrared radiation is ascribed to astronomer William Herschel, who published his results in 1800 before the Royal Society of London. Herschel used a glass prism to refract light from the Sun and detected invisible rays that caused heating beyond the red part of the spectrum, through an increase in the temperature recorded with a thermometer. These "calorific rays" were later termed infrared.

In 1801, German physicist Johann Wilhelm Ritter discovered ultraviolet in an experiment similar to Hershel's, using sunlight and a glass prism. Ritter noted that invisible rays near the violet edge of a solar spectrum dispersed by a triangular prism darkened silver chloride preparations more quickly than did the nearby violet light. Ritter's experiments were an early precursor to what would become photography. Ritter noted that the ultraviolet rays (which at first were called "chemical rays") were capable of causing chemical reactions.

In 1862-4 James Clerk Maxwell developed equations for the electromagnetic field which suggested that waves in the field would travel with a speed that was very close to the known speed of light. Maxwell therefore suggested that visible light (as well as invisible infrared and ultraviolet rays by inference) all consisted of propagating disturbances (or radiation) in the electromagnetic field. Radio waves were first produced deliberately by Heinrich Hertz in 1887, using electrical circuits calculated to produce oscillations at a much lower frequency than that of visible light, following recipes for producing oscillating charges and currents suggested by Maxwell's equations. Hertz also developed ways to detect these waves, and produced and characterized what were later termed radio waves and microwaves.

Wilhelm Röntgen discovered and named X-rays. After experimenting with high voltages applied to an evacuated tube on 8 November 1895, he noticed a fluorescence on a nearby plate of coated glass. In one month, he discovered X-rays' main properties.

The last portion of the EM spectrum to be discovered was associated with radioactivity. Henri Becquerel found that uranium salts caused fogging of an unexposed photographic plate through a covering paper in a manner similar to X-rays, and Marie Curie discovered that only certain elements gave off these rays of energy, soon discovering the intense radiation of radium. The radiation from pitchblende was differentiated into alpha rays (alpha particles) and beta rays (beta particles) by Ernest Rutherford through simple experimentation in 1899, but these proved to be charged particulate types of radiation. However, in 1900 the French scientist Paul Villard discovered a third neutrally charged and especially penetrating type of radiation from radium, and after he described it, Rutherford realized it must be yet a third type of radiation, which in 1903 Rutherford named gamma rays. In 1910 British physicist William Henry Bragg demonstrated that gamma rays are electromagnetic radiation, not particles, and in 1914 Rutherford and Edward Andrade measured their wavelengths, finding that they were similar to X-rays but with shorter wavelengths and higher frequency, although a 'cross-over' between X and gamma rays makes it possible to have X-rays with a higher energy (and hence shorter wavelength) than gamma rays and vice versa. The origin of the ray differentiates them, gamma rays tend to be a natural phenomena originating from the unstable nucleus of an atom and X-rays are electrically generated (and hence man-made) unless they are as a result of bremsstrahlung X-radiation caused by the interaction of fast moving particles (such as beta particles) colliding with certain materials, usually of higher atomic numbers.

Electromagnetic Spectrum

EM radiation (the designation 'radiation' excludes static electric and magnetic and near fields) is classified by wavelength into radio, microwave, infrared, visible, ultraviolet, X-rays and gamma rays. Arbitrary electromagnetic waves can be expressed by Fourier analysis in terms of sinusoidal monochromatic waves, which in turn can each be classified into these regions of the EMR spectrum.

For certain classes of EM waves, the waveform is most usefully treated as *random*, and then spectral analysis must be done by slightly different mathematical techniques appropriate to random or stochastic processes. In such cases, the individual frequency components are represented in terms of their *power* content, and the phase information is not preserved. Such a representation is called the power spectral density of the random process. Random electromagnetic radiation requiring this kind of analysis is, for example, encountered in the interior of stars, and in certain other very wideband forms of radiation such as the Zero point wave field of the electromagnetic vacuum.

The behavior of EM radiation depends on its frequency. Lower frequencies have longer wavelengths, and higher frequencies have shorter wavelengths, and are associated with photons of higher energy. There is no fundamental limit known to these wavelengths or energies, at either end of the spectrum, although photons with energies near the Planck energy or exceeding it (far too high to have ever been observed) will require new physical theories to describe.

Soundwaves are not electromagnetic radiation. At the lower end of the electromagnetic spectrum, about 20 Hz to about 20 kHz, are frequencies that might be considered in the audio range. However, electromagnetic waves cannot be directly perceived by human ears. Sound waves are instead the oscillating compression of molecules. To be heard, electromagnetic radiation must be converted to pressure waves of the fluid in which the ear is located (whether the fluid is air, water or something else).

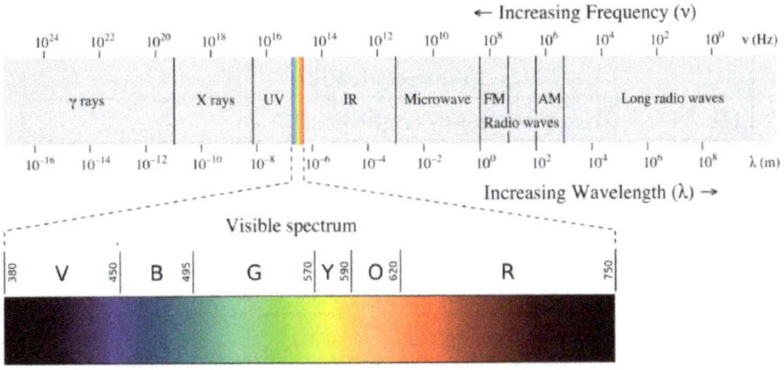

Electromagnetic spectrum with visible light highlighted

CLASS	FREQUENCY	WAVELENGTH	ENERGY
γ	300 EHz	1 pm	1.24 MeV
HX	30 EHz	10 pm	124 keV
	3 EHz	100 pm	12.4 keV
SX	300 PHz	1 nm	1.24 keV
EUV	30 PHz	10 nm	124 eV
NUV	3 PHz	100 nm	12.4 eV
NIR	300 THz	1 μm	1.24 eV
MIR	30 THz	10 μm	124 meV
FIR	3 THz	100 μm	12.4 meV
EHF	300 GHz	1 mm	1.24 meV
SHF	30 GHz	1 cm	124 μeV
UHF	3 GHz	1 dm	12.4 μeV
VHF	300 MHz	1 m	1.24 μeV
HF	30 MHz	10 m	124 neV
MF	3 MHz	100 m	12.4 neV
LF	300 kHz	1 km	1.24 neV
VLF	30 kHz	10 km	124 peV
VF/ULF	3 kHz	100 km	12.4 peV
SLF	300 Hz	1 Mm	1.24 peV
ELF	30 Hz	10 Mm	124 feV
	3 Hz	100 Mm	12.4 feV

Legend	
γ	Gamma rays
HX	Hard X-rays
SX	Soft X-Rays
EUV	Extreme-ultraviolet
NUV	Near-ultraviolet
NIR	Near-infrared
MIR	Mid-infrared
FIR	Far-infrared
EHF	Extremely high frequency (microwaves)
SHF	Super-high frequency (microwaves)
UHF	Ultrahigh frequency (radio waves)
VHF	Very high frequency (radio)
HF	High frequency (radio)
MF	Medium frequency (radio)
LF	Low frequency (radio)
VLF	Very low frequency (radio)
VF	Voice frequency
ULF	Ultra-low frequency (radio)
SLF	Super-low frequency (radio)
ELF	Extremely low frequency(radio)

Interactions as a Function of Frequency

When EM radiation interacts with matter, its behavior changes qualitatively as its frequency changes.

Radio and Microwave

At radio and microwave frequencies, EMR interacts with matter largely as a bulk collection of charges which are spread out over large numbers of affected atoms. In electrical conductors, such induced bulk movement of charges (electric currents) results in absorption of the EMR, or else separations of charges that cause generation of new EMR (effective reflection of the EMR). An example is absorption or emission of radio waves by antennas, or absorption of microwaves by water or other molecules with an electric dipole moment, as for example inside a microwave oven. These interactions produce either electric currents or heat, or both.

Infrared

Like radio and microwave, infrared also is reflected by metals (as is most EMR into the ultraviolet). However, unlike lower-frequency radio and microwave radiation, Infrared

EMR commonly interacts with dipoles present in single molecules, which change as atoms vibrate at the ends of a single chemical bond. It is consequently absorbed by a wide range of substances, causing them to increase in temperature as the vibrations dissipate as heat. The same process, run in reverse, causes bulk substances to radiate in the infrared spontaneously.

Visible Light

As frequency increases into the visible range, photons have enough energy to change the bond structure of some individual molecules. It is not a coincidence that this happens in the "visible range," as the mechanism of vision involves the change in bonding of a single molecule (retinal) which absorbs light in the rhodopsin in the retina of the human eye. Photosynthesis becomes possible in this range as well, for similar reasons, as a single molecule of chlorophyll is excited by a single photon. Animals that detect infrared make use of small packets of water that change temperature, in an essentially thermal process that involves many photons. For this reason, infrared, microwaves and radio waves are thought to damage molecules and biological tissue only by bulk heating, not excitation from single photons of the radiation.

Visible light is able to affect a few molecules with single photons, but usually not in a permanent or damaging way, in the absence of power high enough to increase temperature to damaging levels. However, in plant tissues that conduct photosynthesis, carotenoids act to quench electronically excited chlorophyll produced by visible light in a process called non-photochemical quenching, in order to prevent reactions that would otherwise interfere with photosynthesis at high light levels. Limited evidence indicate that some reactive oxygen species are created by visible light in skin, and that these may have some role in photoaging, in the same manner as ultraviolet A.

Ultraviolet

As frequency increases into the ultraviolet, photons now carry enough energy (about three electron volts or more) to excite certain doubly bonded molecules into permanent chemical rearrangement. In DNA, this causes lasting damage. DNA is also indirectly damaged by reactive oxygen species produced by ultraviolet A (UVA), which has energy too low to damage DNA directly. This is why ultraviolet at all wavelengths can damage DNA, and is capable of causing cancer, and (for UVB) skin burns (sunburn) that are far worse than would be produced by simple heating (temperature increase) effects. This property of causing molecular damage that is out of proportion to heating effects, is characteristic of all EMR with frequencies at the visible light range and above. These properties of high-frequency EMR are due to quantum effects that permanently damage materials and tissues at the molecular level.

At the higher end of the ultraviolet range, the energy of photons becomes large enough to impart enough energy to electrons to cause them to be liberated from the atom, in a process called photoionisation. The energy required for this is always larger than about 10 electron volts (eV) corresponding with wavelengths smaller than 124 nm (some sources suggest a more realistic cutoff of 33 eV, which is the energy required to ionize water). This high end of the ultraviolet spectrum with energies in the approximate ionization range, is sometimes called "extreme UV." Ionizing UV is strongly filtered by the Earth's atmosphere).

X-rays and Gamma Rays

Electromagnetic radiation composed of photons that carry minimum-ionization energy, or more, (which includes the entire spectrum with shorter wavelengths), is therefore termed ionizing radiation. (Many other kinds of ionizing radiation are made of non-EM particles). Electromagnetic-type ionizing radiation extends from the extreme ultraviolet to all higher frequencies and shorter wavelengths, which means that all X-rays and gamma rays qualify. These are capable of the most severe types of molecular damage, which can happen in biology to any type of biomolecule, including mutation and cancer, and often at great depths below the skin, since the higher end of the X-ray spectrum, and all of the gamma ray spectrum, penetrate matter.

Atmosphere and Magnetosphere

Most UV and X-rays are blocked by absorption first from molecular nitrogen, and then (for wavelengths in the upper UV) from the electronic excitation of dioxygen and finally ozone at the mid-range of UV. Only 30% of the Sun's ultraviolet light reaches the ground, and almost all of this is well transmitted.

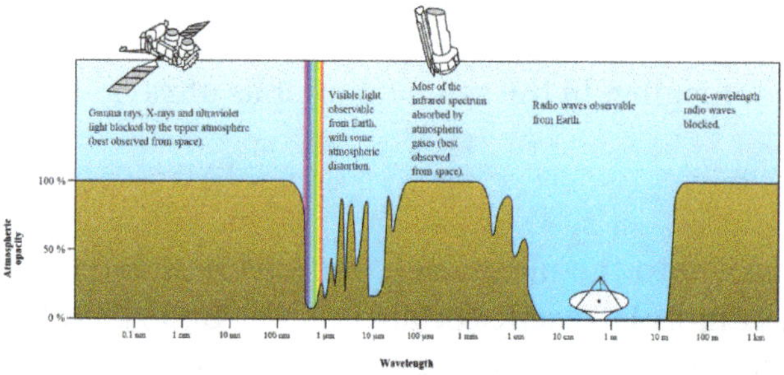

Rough plot of Earth's atmospheric absorption and scattering (or opacity) of various wavelengths of electromagnetic radiation

Visible light is well transmitted in air, as it is not energetic enough to excite nitrogen, oxygen, or ozone, but too energetic to excite molecular vibrational frequencies of water vapor.

Absorption bands in the infrared are due to modes of vibrational excitation in water vapor. However, at energies too low to excite water vapor, the atmosphere becomes transparent again, allowing free transmission of most microwave and radio waves.

Finally, at radio wavelengths longer than 10 meters or so (about 30 MHz), the air in the lower atmosphere remains transparent to radio, but plasma in certain layers of the ionosphere begins to interact with radio waves. This property allows some longer wavelengths (100 meters or 3 MHz) to be reflected and results in shortwave radio beyond line-of-sight. However, certain ionospheric effects begin to block incoming radiowaves from space, when their frequency is less than about 10 MHz (wavelength longer than about 30 meters).

Types and Sources, Classed by Spectral Band

Radio Waves

Radio waves have the least amount of energy and the lowest frequency. When radio waves impinge upon a conductor, they couple to the conductor, travel along it and induce an electric current on the conductor surface by moving the electrons of the conducting material in correlated bunches of charge. Such effects can cover macroscopic distances in conductors (such as radio antennas), since the wavelength of radiowaves is long.

Microwaves

Microwaves are a form of electromagnetic radiation with wavelengths ranging from as long as one meter to as short as one millimeter; with frequencies between 300 MHz (0.3 GHz) and 300 GHz.

Visible Light

Natural sources produce EM radiation across the spectrum. EM radiation with a wavelength between approximately 400 nm and 700 nm is directly detected by the human eye and perceived as visible light. Other wavelengths, especially nearby infrared (longer than 700 nm) and ultraviolet (shorter than 400 nm) are also sometimes referred to as light.

Thermal Radiation and Electromagnetic Radiation as a Form of Heat

The basic structure of matter involves charged particles bound together. When electromagnetic radiation impinges on matter, it causes the charged particles to oscillate and gain energy. The ultimate fate of this energy depends on the context. It could be immediately re-radiated and appear as scattered, reflected, or transmitted radiation. It may get dissipated into other microscopic motions within the matter, coming to thermal equilibrium and manifesting itself as thermal energy in the material. With a few exceptions related to high-energy photons (such as fluorescence, harmonic generation, photochemical reactions, the photovoltaic effect for ionizing radiations at far ultraviolet, X-ray and gamma radiation), absorbed electromagnetic radiation simply deposits its energy by heating the material. This happens for infrared, microwave and radio wave radiation. Intense radio waves can thermally burn living tissue and can cook food. In addition to infrared lasers, sufficiently intense visible and ultraviolet lasers can easily set paper afire.

Ionizing radiation creates high-speed electrons in a material and breaks chemical bonds, but after these electrons collide many times with other atoms eventually most of the energy becomes thermal energy all in a tiny fraction of a second. This process makes ionizing radiation far more dangerous per unit of energy than non-ionizing radiation. This caveat also applies to UV, even though almost all of it is not ionizing, because UV can damage molecules due to electronic excitation, which is far greater per unit energy than heating effects.

Infrared radiation in the spectral distribution of a black body is usually considered a form of heat, since it has an equivalent temperature and is associated with an entropy change per unit of thermal energy. However, "heat" is a technical term in physics and thermodynamics and is often confused with thermal energy. Any type of electromagnetic energy can be transformed into thermal energy in interaction with matter. Thus, *any* electromagnetic radiation can "heat" (in the sense of increase the thermal energy termperature of) a material, when it is absorbed.

The inverse or time-reversed process of absorption is thermal radiation. Much of the thermal energy in matter consists of random motion of charged particles, and this energy can be radiated away from the matter. The resulting radiation may subsequently be absorbed by another piece of matter, with the deposited energy heating the material.

The electromagnetic radiation in an opaque cavity at thermal equilibrium is effectively a form of thermal energy, having maximum radiation entropy.

Biological Effects

Bioelectromagnetics is the study of the interactions and effects of EM radiation on living organisms. The effects of electromagnetic radiation upon living cells, including those in humans, depends upon the radiation's power and frequency. For low-frequency radiation (radio waves to visible light) the best-understood effects are those due to radiation power alone, acting through heating when radiation is absorbed. For these thermal effects, frequency is important only as it affects penetration into the organism (for example, microwaves penetrate better than infrared). Initially, it was believed that low frequency fields that were too weak to cause significant heating could not possibly have any biological effect.

Despite this opinion among researchers, evidence has accumulated that supports the existence of complex biological effects of weaker *non-thermal* electromagnetic fields, (including weak ELF magnetic fields, although the latter does not strictly qualify as EM radiation), and modulated RF and microwave fields. Fundamental mechanisms of the interaction between biological material and electromagnetic fields at non-thermal levels are not fully understood.

The World Health Organization has classified radio frequency electromagnetic radiation as Group 2B - possibly carcinogenic. This group contains possible carcinogens that have weaker evidence, at the same level as coffee and automobile exhaust. For example, epidemiological studies looking for a relationship between cell phone use and brain cancer development, have been largely inconclusive, save to demonstrate that the effect, if it exists, cannot be a large one.

At higher frequencies (visible and beyond), the effects of individual photons begin to become important, as these now have enough energy individually to directly or indirectly damage biological molecules. All UV frequences have been classed as Group 1 carcinogens by the World Health Organization. Ultraviolet radiation from sun exposure is the primary cause of skin cancer.

Thus, at UV frequencies and higher (and probably somewhat also in the visible range), electromagnetic radiation does more damage to biological systems than simple heating predicts. This is most obvious in the "far" (or "extreme") ultraviolet. UV, with X-ray and gamma radiation, are referred to as ionizing radiation due to the ability of photons of this radiation to produce ions and free radicals in materials (including living tissue). Since such radiation can severely damage life at energy levels that produce little heating, it is considered far more dangerous (in terms of damage-produced per unit of energy, or power) than the rest of the electromagnetic spectrum.

Derivation from Electromagnetic Theory

Electromagnetic waves were predicted by the classical laws of electricity and magnetism, known as Maxwell's equations. Inspection of Maxwell's equations without sources (charges or currents) results in nontrivial solutions of changing electric and magnetic fields. Beginning with Maxwell's equations in free space:

$$\nabla \cdot \mathbf{E} = 0$$

$$\nabla \times \mathbf{E} = -\frac{\partial \mathbf{B}}{\partial t}$$

$$\nabla \cdot \mathbf{B} = 0$$

$$\nabla \times \mathbf{B} = \mu_0 \epsilon_0 \frac{\partial \mathbf{E}}{\partial t}$$

where

∇ is a vector differential operator.

One solution,

$$\mathbf{E} = \mathbf{B} = \mathbf{0},$$

is trivial.

For a more useful solution, we utilize vector identities, which work for any vector, as follows:

$$\nabla \times (\nabla \times \mathbf{A}) = \nabla(\nabla \cdot \mathbf{A}) - \nabla^2 \mathbf{A}$$

The curl of equation (2):

$$\nabla \times (\nabla \times \mathbf{E}) = \nabla \times \left(-\frac{\partial \mathbf{B}}{\partial t} \right)$$

Evaluating the left hand side:

$$\nabla \times (\nabla \times \mathbf{E}) = \nabla(\nabla \cdot \mathbf{E}) - \nabla^2 \mathbf{E} = -\nabla^2 \mathbf{E}$$

simplifying the above equation.

Evaluating the right hand side:

$$\nabla \times \left(-\frac{\partial \mathbf{B}}{\partial t} \right) = -\frac{\partial}{\partial t}(\nabla \times \mathbf{B}) = -\mu_0 \epsilon_0 \frac{\partial^2 \mathbf{E}}{\partial t^2}$$

Equations are equal, so this results in a vector-valued differential equation for the electric field, namely

$$\nabla^2 \mathbf{E} = \mu_0 \epsilon_0 \frac{\partial^2 \mathbf{E}}{\partial t^2}$$

Applying a similar pattern results in similar differential equation for the magnetic field:

$$\nabla^2 \mathbf{B} = \mu_0 \epsilon_0 \frac{\partial^2 \mathbf{B}}{\partial t^2}.$$

These differential equations are equivalent to the wave equation:

$$\nabla^2 f = \frac{1}{c_0^2} \frac{\partial^2 f}{\partial t^2}$$

where

c_o is the speed of the wave in free space and

f describes a displacement

Or more simply:

$$\Box f = 0$$

where $\square$ is d'Alembertian:

$$\square = \nabla^2 - \frac{1}{c_0^2}\frac{\partial^2}{\partial t^2} = \frac{\partial^2}{\partial x^2} + \frac{\partial^2}{\partial y^2} + \frac{\partial^2}{\partial z^2} - \frac{1}{c_0^2}\frac{\partial^2}{\partial t^2}$$

In the case of the electric and magnetic fields, the speed is:

$$c_0 = \frac{1}{\sqrt{\mu_0 \epsilon_0}}$$

This is the speed of light in vacuum. Maxwell's equations unified the vacuum permittivity ϵ_0, the vacuum permeability μ_0, and the speed of light itself, c_0. This relationship had been discovered by Wilhelm Eduard Weber and Rudolf Kohlrausch prior to the development of Maxwell's electrodynamics, however Maxwell was the first to produce a field theory consistent with waves traveling at the speed of light.

These are only two equations versus the original four, so more information pertains to these waves hidden within Maxwell's equations. A generic vector wave for the electric field.

$$\mathbf{E} = \mathbf{E}_0 f\left(\hat{\mathbf{k}} \cdot \mathbf{x} - c_0 t\right)$$

Here, $\mathbf{E}_0$ is the constant amplitude, f is any second differentiable function, $\hat{\mathbf{k}}$ is a unit vector in the direction of propagation, and $\mathbf{x}$ is a position vector. $f\left(\hat{\mathbf{k}} \cdot \mathbf{x} - c_0 t\right)$ is a generic solution to the wave equation. In other words,

$$\nabla^2 f\left(\hat{\mathbf{k}} \cdot \mathbf{x} - c_0 t\right) = \frac{1}{c_0^2}\frac{\partial^2}{\partial t^2} f\left(\hat{\mathbf{k}} \cdot \mathbf{x} - c_0 t\right),$$

for a generic wave traveling in the $\hat{\mathbf{k}}$ direction.

This form will satisfy the wave equation.

$$\nabla \cdot \mathbf{E} = \hat{\mathbf{k}} \cdot \mathbf{E}_0 f'\left(\hat{\mathbf{k}} \cdot \mathbf{x} - c_0 t\right) = 0$$

$$\mathbf{E} \cdot \hat{\mathbf{k}} = 0$$

The first of Maxwell's equations implies that the electric field is orthogonal to the direction the wave propagates.

$$\nabla \times \mathbf{E} = \hat{\mathbf{k}} \times \mathbf{E}_0 f'\left(\hat{\mathbf{k}} \cdot \mathbf{x} - c_0 t\right) = -\frac{\partial \mathbf{B}}{\partial t}$$

$$\mathbf{B} = \frac{1}{c_0}\hat{\mathbf{k}} \times \mathbf{E}$$

The second of Maxwell's equations yields the magnetic field. The remaining equations will be satisfied by this choice of $\mathbf{E}, \mathbf{B}$.

The electric and magnetic field waves in the far-field travel at the speed of light. They have a special restricted orientation and proportional magnitudes, $E_0 = c_0 B_0$, which can be seen immediately from the Poynting vector. The electric field, magnetic field, and direction of wave propagation are all orthogonal, and the wave propagates in the same direction as $\mathbf{E} \times \mathbf{B}$. Also, E and B far-fields in free space, which as wave solutions depend primarily on these two Maxwell equations, are in-phase with each other. This is guaranteed since the generic wave solution is first order in both space and time, and the curl operator on one side of these equations results in first-order spatial derivatives of the wave solution, while the time-derivative on the other side of the equations, which gives the other field, is first order in time, resulting in the same phase shift for both fields in each mathematical operation.

From the viewpoint of an electromagnetic wave traveling forward, the electric field might be oscillating up and down, while the magnetic field oscillates right and left. This picture can be rotated with the electric field oscillating right and left and the magnetic field oscillating down and up. This is a different solution that is traveling in the same direction. This arbitrariness in the orientation with respect to propagation direction is known as polarization. On a quantum level, it is described as photon polarization. The direction of the polarization is defined as the direction of the electric field.

More general forms of the second-order wave equations given above are available, allowing for both non-vacuum propagation media and sources. Many competing derivations exist, all with varying levels of approximation and intended applications. One very general example is a form of the electric field equation, which was factorized into a pair of explicitly directional wave equations, and then efficiently reduced into a single uni-directional wave equation by means of a simple slow-evolution approximation.

Maxwell's Equations

Maxwell's equations are a set of partial differential equations that, together with the Lorentz force law, form the foundation of classical electromagnetism, classical optics, and electric circuits. They underpin all electric, optical and radio technologies such as power generation, electric motors, wireless communication, cameras, televisions, computers etc. Maxwell's equations describe how electric and magnetic fields are generated by charges, currents and changes of each other. One important consequence of the equations is the demonstration of how fluctuating electric and magnetic fields can propagate at the speed of light. This electromagnetic radiation manifests itself in manifold ways from radio waves to light and X- or γ-rays. The equations are named after the physicist and mathematician James Clerk Maxwell, who between 1861 and 1862 published an early form of the equations, and first proposed that light is an electromagnetic phenomenon.

The equations have two major variants. The microscopic Maxwell equations have universal applicability but are unwieldy for common calculations. They relate the electric and magnetic fields to total charge and total current, including the complicated charges and currents in materials at the atomic scale. The "macroscopic" Maxwell equations define two new auxiliary fields that describe the large-scale behaviour of matter without having to consider atomic scale details. However, their use requires experimentally determining parameters for a phenomenological description of the electromagnetic response of materials.

The term "Maxwell's equations" is often used for equivalent alternative formulations. Versions of Maxwell's equations based on the electric and magnetic potentials are preferred for explicitly solving the equations as a boundary value problem, analytical mechanics, or for use in quantum mechanics. The space-time formulations (i.e., on space-time rather than space and time separately), are commonly used in high energy and gravitational physics because they make the compatibility of the equations with special and general relativity manifest. In fact, historically, Einstein developed special and general relativity to accommodate the absolute speed of light that drops out of the Maxwell equations with the principle that only relative movement has physical consequences.

Since the mid-20th century, it has been understood that Maxwell's equations are not exact but are a classical field theory approximation of some aspects of the fundamental theory of quantum electrodynamics, although some quantum features, such as quantum entanglement, are completely absent and in no way approximated. (For example, quantum cryptography has no approximate version in Maxwell theory.) In many situations, though, deviations from Maxwell's equations are immeasurably small. Exceptions include nonclassical light, photon-photon scattering, quantum optics, and many other phenomena related to photons or virtual photons.

Formulation in Terms of Electric and Magnetic Fields (Microscopic or in Vacuum Version)

In the electric and magnetic field formulation there are four equations. The two inhomogeneous equations describe how the fields vary in space due to sources. Gauss's law describes how electric fields emanate from electric charges. Gauss's law for magnetism describes magnetic fields as closed field lines *not due to magnetic monopoles*. The two homogeneous equations describe how the fields "circulate" around their respective sources. Ampère's law with Maxwell's addition describes how the magnetic field "circulates" around electric currents and time varying electric fields, while Faraday's law describes how the electric field "circulates" around time varying magnetic fields.

A separate law of nature, the Lorentz force law, describes how the electric and magnetic field act on charged particles and currents. A version of this law was included in the original equations by Maxwell but, by convention, is no longer.

The precise formulation of Maxwell's equations depends on the precise definition of the quantities involved. Conventions differ with the unit systems, because various definitions and dimensions are changed by absorbing dimensionful factors like the speed of light c. This makes constants come out differently. The most common form is based on conventions used when quantities measured using SI units, but other commonly used conventions are used with other units including Gaussian units based on the cgs system, Lorentz–Heaviside units (used mainly in particle physics), and Planck units (used in theoretical physics).

The vector calculus formulation below has become standard. It is mathematically much more convenient than Maxwell's original 20 equations and is due to Oliver Heaviside The differential and integral equations formulations are mathematically equivalent and are both useful. The integral formulation relates fields within a region of space to fields on the boundary and can often be used to simplify and directly calculate fields from symmetric distributions of charges and currents. On the other hand, the differential equations are purely *local* and are a more natural starting point for calculating the fields in more complicated (less symmetric) situations, for example using finite element analysis.

Formulation in SI Units Convention

Name	Integral equations	Differential equations	Meaning
Gauss's law	$\oiint_{\partial\Omega} E.\,dS = \frac{1}{\varepsilon_0}\iiint_{\Omega} p dV$	$\nabla.\,E = \frac{p}{\varepsilon_0}$	The electric flux leaving a volume is proportional to the charge inside.
Gauss's law for magnetism	$\oiint_{\partial\Omega} B.\,dS = 0$	$\nabla.\,B = 0$	There are no magnetic monopoles; the total magnetic flux through a closed surface is zero.
Maxwell–Faraday equation(Faraday's law of induction)	$\oint_{\partial\Sigma} E.\,d\ell = -\frac{d}{dt}\iint_{\Sigma} B.\,dS$	$\nabla\times E = -\frac{\partial B}{\partial t}$	The voltage induced in a closed circuit is proportional to the rate of change of the magnetic flux it encloses.
Ampère's circuital law (with Maxwell's addition)	$\oint_{\partial\Sigma} B.\,d\ell = \mu_0\iint_{\Sigma} J.\,dS + \mu_0\varepsilon_0\frac{d}{dt}\iint_{\Sigma} E.ds$	$\nabla.B = \mu_0\left(J + \varepsilon_0\frac{\partial E}{\partial t}\right)$	The magnetic field induced around a closed loop is proportional to the electric current plus displacement current (rate of change of electric field) it encloses.

Formulation in Gaussian Units Convention

Gaussian units are a popular system of units, that are part of the centimetre–gram–second system of units (cgs). When using cgs units it is conventional to use a slightly different definition of electric field $E_{cgs} = c^{-1} E_{SI}$. This implies that the modified electric and magnetic field have the same units (in the SI convention this is not the case making dimensional analysis of the equations different: e.g. for an electromagnetic wave in vacuum $|\mathbf{E}|_{SI} = |c|_{SI} |\mathbf{B}|_{SI}$,). The CGS system uses a unit of charge defined in such a way that the permittivity of the vacuum $\varepsilon_0 = 1/4\pi c$, hence $\mu_0 = 4\pi/c$. These units are sometimes preferred over SI units in the context of special relativity, since when using them, the components of the electromagnetic tensor, the Lorentz covariant object describing the electromagnetic field, have the same unit without constant factors. Using these different conventions, the Maxwell equations become:

Name	Integral equations	Differential equations	Meaning
Gauss's law	$\oiint_{\partial\Omega} E.\, dS = 4\pi \iiint_{\Omega} p dV$	$\nabla \cdot \mathbf{E} = 4\pi p$	The electric flux leaving a volume is proportional to the charge inside.
Gauss's law for magnetism	$\oiint_{\partial\Omega} B.\, dS = 0$	$\nabla \cdot \mathbf{B} = 0$	There are no magnetic monopoles; the total magnetic flux through a closed surface is zero.
Maxwell–Faraday equation (Faraday's law of induction)	$\oint_{\partial\Sigma} E.d\ell = -\frac{1}{c}\frac{d}{dt}\iint_{\Sigma} B.dS$	$\nabla \times \mathbf{E} = -\frac{1}{c}\frac{\partial \mathbf{B}}{\partial t}$	The voltage induced in a closed circuit is proportional to the rate of change of the magnetic flux it encloses.
Ampère's law (with Maxwell's extension)	$\oint_{\partial\Omega} B.d\ell = \iint_{\Sigma}$ $\frac{1}{c}\left(4\pi\iint_{\Sigma} J.\, dS + \frac{d}{dt}\iint_{\Sigma} E.\, dS\right)$	$\iint_{\Sigma}$ $\nabla \times \mathbf{B} = \frac{1}{c}\left(4\pi \mathbf{J} + \frac{\partial \mathbf{E}}{\partial t}\right)$	The magnetic field integrated around a closed loop is proportional to the electric current plus displacement current (rate of change of electric field) it encloses.

Key to the Notation

Symbols in bold represent vector quantities, and symbols in *italics* represent scalar quantities, unless otherwise indicated.

The equations introduce the electric field, E, a vector field, and the magnetic field, B, a pseudovector field, each generally having a time and location dependence. The sources are:

- the electric charge density (charge per unit volume), ρ, and
- the electric current density (current per unit area), J.

The universal constants appearing in the equations are

- the permittivity of free space, ε_o, and
- the permeability of free space, μ_o.

Differential Equations

In the differential equations,

- the nabla symbol, ∇, denotes the three-dimensional gradient operator,
- the $\nabla\cdot$ symbol denotes the divergence operator,
- the $\nabla\times$ symbol denotes the curl operator.

Integral Equations

In the integral equations,

- Ω is any fixed volume with closed boundary surface $\partial\Omega$, and
- Σ is any fixed surface with closed boundary curve $\partial\Sigma$,

Here a *fixed* volume or surface means that it does not change over time. The equations are correct, complete and a little easier to interpret with time-independent surfaces. However, since the surface is time-independent, we can bring the differentiation under the integral sign in Faraday's law:

$$\frac{d}{dt}\iint_{\Sigma} B.\,dS = \iint_{\Sigma} \frac{\partial B}{\partial t}.\,dS,$$

The Maxwell's equations can be formulated with possibly time dependent surfaces and volumes by substituting the lefthand side with the righthand side in the integral equation version of the Maxwell equations.

- $\oiint_{\partial\Omega}$ is a surface integral over the surface $\partial\Omega$, (the loop indicates that the boundary surface is closed)
- $\iiint_{\Omega}$ is a volume integral over the volume Ω,
- $\oint_{\partial\Sigma}$ is a line integral around the curve $\partial\Sigma$ (the loop indicates that the boundary curve is closed).
- $\iint_{\Sigma}$ is a surface integral over the surface Σ,

- The volume integral over Ω of the total charge density ρ, is the *total* electric charge Q contained in Ω:

$$Q = \iiint_\Omega p dV$$

where dV is the volume element.

- The *net* electric current I is the surface integral of the electric current density J passing through a fixed surface, Σ:

$$I = \iint_\Sigma J.\, dS,$$

where dS denotes the vector element of surface area, S, normal to surface, Σ. (Vector area is also denoted by A rather than S, but this conflicts with the magnetic potential, a separate vector field).

Relationship between Differential and Integral Formulations

The equivalence of the differential and integral formulations are a consequence of the Gauss divergence theorem and the Kelvin–Stokes theorem.

Flux and Divergence

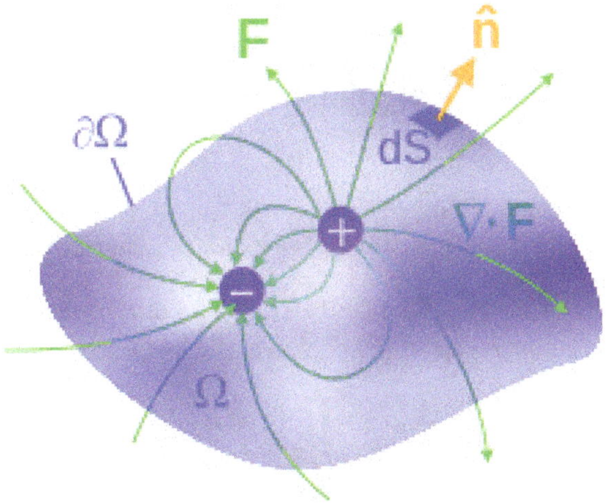

Volume Ω and its closed boundary $\partial\Omega$, containing (respectively enclosing) a source (+) and sink (−) of a vector field F. Here, F could be the E field with source electric charges, but *not* the B field which has no magnetic charges as shown. The outward unit normal is n.

The "sources of the fields" (i.e. their divergence) can be determined from the surface integrals of the fields through the closed surface $\partial\Omega$. E.g. the electric flux is

$$\oiint_{\partial\Omega} E.\, dS = \iiint_\Omega \nabla . E dV$$

where the last equality uses the Gauss divergence theorem. Using the integral version of Gauss's equation we can rewrite this to

$$\iiint_{\Omega}\left(\nabla . E-\frac{\rho}{\varepsilon_0}\right)dV=0$$

Since Ω can be chosen arbitrarily, e.g. as an arbitrary small ball with arbitrary center, this implies that the integrand must be zero, which is the differential equations formulation of Gauss equation up to a trivial rearrangement. Gauss's law for magnetism in differential equations form follows likewise from the integral form by rewriting the magnetic flux

$$\oiint_{\partial\Omega} B . dS=\iiint_{\Omega}\nabla . B dV=0$$

Circulation and Curl

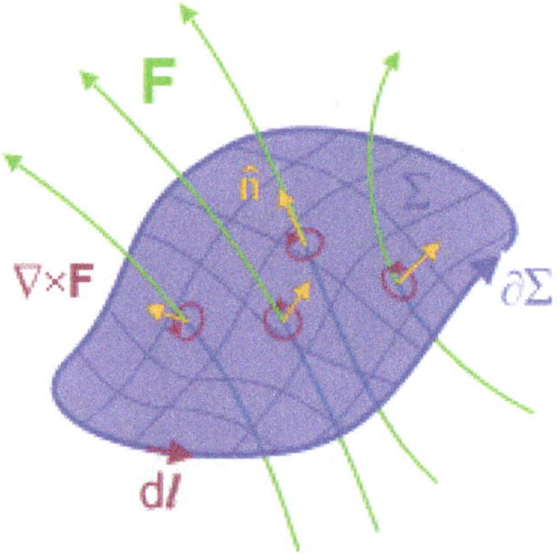

Surface Σ with closed boundary $\partial\Sigma$. F could be the E or B fields. Again, n is the unit normal. (The curl of a vector field doesn't literally look like the "circulations", this is a heuristic depiction).

The "circulation of the fields" (i.e. their curls) can be determined from the line integrals of the fields around the closed curve $\partial\Sigma$. E.g. for the magnetic field

$$\oint_{\partial\Sigma} B . d\ell=\iint_{\Sigma}(\nabla\times B) . dS,$$

where we used the Kelvin-Stokes theorem. Using the modified Ampere law in integral form and the writing the time derivative of the flux as the surface integral of the partial time derivative of E we conclude that

$$\iint_{\Sigma}\left(\nabla\times B-\mu_0\left(J+\varepsilon_0\frac{\partial E}{\partial t}\right)\right). dS=0$$

Since Σ can be chosen arbitrarily, e.g. as an arbitrary small, arbitrary oriented, and arbitrary centered disk, we conclude that the integrand must be zero. This is Ampere's modified law in differential equations form up to a trivial rearrangement. Likewise, the Faraday law in differential equations form follows from rewriting the integral form using the Kelvin-Stokes theorem.

The line integrals and curls are analogous to quantities in classical fluid dynamics: the circulation of a fluid is the line integral of the fluid's flow velocity field around a closed loop, and the vorticity of the fluid is the curl of the velocity field.

Conceptual Descriptions

Gauss's Law

Gauss's law describes the relationship between a static electric field and the electric charges that cause it: The static electric field points away from positive charges and towards negative charges. In the field line description, electric field lines begin only at positive electric charges and end only at negative electric charges. 'Counting' the number of field lines passing through a closed surface, therefore, yields the total charge (including bound charge due to polarization of material) enclosed by that surface divided by dielectricity of free space (the vacuum permittivity). More technically, it relates the electric flux through any hypothetical closed "Gaussian surface" to the enclosed electric charge.

Gauss's Law for Magnetism

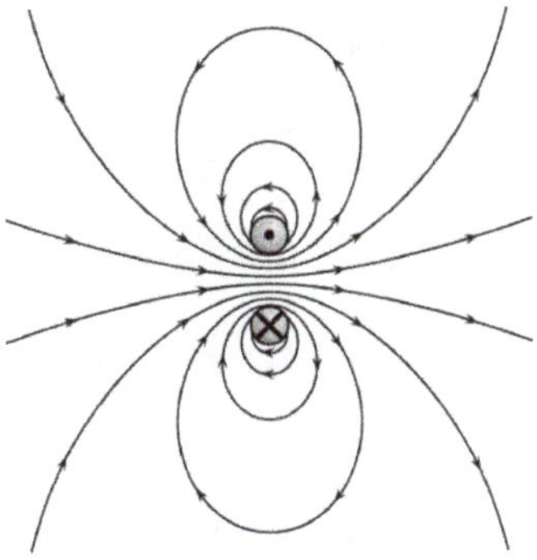

Gauss's law for magnetism: magnetic field lines never begin nor end but form loops or extend to infinity as shown here with the magnetic field due to a ring of current.

Gauss's law for magnetism states that there are no "magnetic charges" (also called magnetic monopoles), analogous to electric charges. Instead, the magnetic field due to materials is generated by a configuration called a dipole. Magnetic dipoles are best represented as loops of current but resemble positive and negative 'magnetic charges', inseparably bound together, having no net 'magnetic charge'. In terms of field lines, this equation states that magnetic field lines neither begin nor end but make loops or extend to infinity and back. In other words, any magnetic field line that enters a given volume must somewhere exit that volume. Equivalent technical statements are that the sum total magnetic flux through any Gaussian surface is zero, or that the magnetic field is a solenoidal vector field.

Faraday's Law

In a geomagnetic storm, a surge in the flux of charged particles temporarily alters Earth's magnetic field, which induces electric fields in Earth's atmosphere, thus causing surges in electrical power grids. Artist's rendition; sizes are not to scale.

The Maxwell-Faraday's equation version of Faraday's law describes how a time varying magnetic field creates ("induces") an electric field. This dynamically induced electric field has closed field lines just as the magnetic field, if not superposed by a static (charge induced) electric field. This aspect of electromagnetic induction is the operating principle behind many electric generators: for example, a rotating bar magnet creates a changing magnetic field, which in turn generates an electric field in a nearby wire.

Ampère's Law with Maxwell's Addition

Magnetic core memory (1954) is an application of Ampère's law. Each core stores one bit of data.

Ampère's law with Maxwell's addition states that magnetic fields can be generated in two ways: by electric current (this was the original "Ampère's law") and by changing electric fields (this was "Maxwell's addition").

Maxwell's addition to Ampère's law is particularly important: it makes the set of equations mathematically consistent for non static fields, without changing the laws of Ampere and Gauss for static fields. However, as a consequence, it predicts that a changing magnetic field induces an electric field and vice versa. Therefore, these equations allow self-sustaining "electromagnetic waves" to travel through empty space.

The speed calculated for electromagnetic waves, which could be predicted from experiments on charges and currents, exactly matches the speed of light; indeed, light *is* one form of electromagnetic radiation (as are X-rays, radio waves, and others). Maxwell understood the connection between electromagnetic waves and light in 1861, thereby unifying the theories of electromagnetism and optics.

Charge Conservation

The lefthand side of the modified Ampere's law has zero divergence by the div-curl-identity. Therefore, the right handside, Gauss's law and interchanging derivatives give

$$0 = \nabla.\nabla \times B = \mu_0 \left(\nabla.J + \varepsilon_0 \frac{\partial}{\partial t} \nabla.\mathrm{E} \right) = \mu_0 \left(\nabla.J + \frac{\partial \rho}{\partial t} \right)$$

i.e.

$$\frac{\partial \mathrm{p}}{\partial \mathrm{t}} + \nabla.\,\mathrm{J} = 0$$

By the Gauss divergence theorem that means that the rate of change of the charge in a fixed volume equals the current flowing in or out of the boundary

$$\frac{\mathrm{d}}{\mathrm{dt}} \mathrm{Q}_{\Omega} = \frac{\mathrm{d}}{\mathrm{dt}} \iiint_{\Omega} \mathrm{pdV} = -\oiint_{\partial\Omega} \mathrm{J}.\,\mathrm{dS} = -\mathrm{I}_{\partial\Omega}$$

In particular, in an isolated system the total charge is conserved.

Vacuum Equations, Electromagnetic Waves and Speed of Light

In a region with no charges ($\rho = 0$) and no currents ($J = 0$), such as in a vacuum, Maxwell's equations reduce to:

$$\nabla.\,E = 0 \qquad \nabla \times E = -\frac{\partial B}{\partial t},$$

$$\nabla.\,B = 0 \qquad \nabla \times B = \frac{1}{c^2} \frac{\partial E}{\partial t}.$$

Taking the curl ($\nabla\times$) of the curl equations, and using the curl of the curl identity $\nabla \times (\nabla \times X) = \nabla(\nabla \cdot X) - \nabla^2 X$ we obtain the wave equations

$$\frac{1}{c^2} \frac{\partial^2 E}{\partial t^2} - \nabla^2 E = 0$$

$$\frac{1}{c^2} \frac{\partial^2 B}{\partial t^2} - \nabla^2 B = 0$$

which identify

$$c = \frac{1}{\sqrt{\mu_0 \in_0}} = 2.99792485 \times 10^8 \, m s^{-1}$$

with the speed of light in free space. In materials with relative permittivity, ε_r, and relative permeability, μ_r, the phase velocity of light becomes

$$\upsilon_p = \frac{1}{\sqrt{\mu_0 \mu_r \varepsilon_0 \varepsilon_r}}$$

which is usually less than c.

In addition, E and B are mutually perpendicular to each other and the direction of wave propagation, and are in phase with each other. A sinusoidal plane wave is one special solution of these equations. Maxwell's equations explain how these waves can physically propagate through space. The changing magnetic field creates a changing electric field through Faraday's law. In turn, that electric field creates a changing magnetic field through Maxwell's addition to Ampère's law. This perpetual cycle allows these waves, now known as electromagnetic radiation, to move through space at velocity c.

Macroscopic Formulation

The *microscopic* variant of Maxwell's equation is the version given above. It expresses the electric E field and the magnetic B field in terms of the *total charge* and *total current* present, including the charges and currents at the atomic level. The "microscopic" form is sometimes called the "general" form of Maxwell's equations. The macroscopic variant of Maxwell's equation is equally general, however, with the difference being one of bookkeeping.

The "microscopic" variant is sometimes called "Maxwell's equations in a vacuum". This refers to the fact that the material medium is not built into the structure of the equation; it does not mean that space is empty of charge or current. They are also known as the "Maxwell-Lorentz equations". Lorentz tried to use these equations to predict the macroscopic properties of bulk matter from the physical behavior of its microscopic constituents.

"Maxwell's macroscopic equations", also known as Maxwell's equations in matter, are more similar to those that Maxwell introduced himself.

Name	Integral equations (SI convention)	Differential equations (SI convention)	Differential equations (Gaussian convention)
Gauss's law	$\oiint_{\partial\Omega} \mathrm{D}.\,d\mathrm{S} = \iiint_{\Omega} \rho_f dV$	$\nabla.\mathrm{D} = \rho_f$	$\nabla.\mathrm{D} = 4\pi\rho_f$
Gauss's law for magnetism	$\oiint_{\partial\Omega} \mathrm{B}.\,d\mathrm{S} = 0$	$\nabla.\mathrm{B} = 0$	$\nabla.\mathrm{B} = 0$
Maxwell–Faraday equation (Faraday's law of induction)	$\oint_{\partial\Sigma} \mathrm{E}.d\ell = -\frac{d}{dt}\iint_{\Sigma} \mathrm{B}.\,d\mathrm{S}$	$\nabla \times \mathrm{E} = -\frac{\partial \mathrm{B}}{\partial t}$	$\nabla \times \mathrm{E} = -\frac{1}{c}\frac{\partial \mathrm{B}}{\partial t}$
Ampère's circuital law (with Maxwell's addition)	$\oint_{\partial\Sigma} \mathrm{H}.\,d\ell = \iint_{\Sigma} \mathrm{J}_f.\,d\mathrm{S} + \frac{d}{dt}\iint_{\Sigma} \mathrm{D}.\,d\mathrm{S}$	$\nabla \times \mathrm{H} = \mathrm{J}_f + \frac{\partial \mathrm{D}}{\partial t}$	$\nabla \times \mathrm{H} = \frac{1}{c}\left(4\pi \mathrm{J}_f + \frac{\partial \mathrm{D}}{\partial t}\right)$

Unlike the "microscopic" equations, the "macroscopic" equations separate out the bound charge Q_b and bound current I_b to obtain equations that depend only on the free charges Q_f and currents I_f. This factorization can be made by splitting the total electric charge and current as follows:

$$Q = Q_f + Q_b = \iiint_{\Omega} (\rho_f + \rho_b)\, dV = \iiint_{\Omega} \rho\, dV$$

$$I = I_f + I_b = \iint_{\Sigma} (J_f + J_b) . dS = \iint_{\Sigma} J. dS$$

Correspondingly, the total current density J splits into free J_f and bound J_b components, and similarly the total charge density ρ splits into free ρ_f and bound ρ_b parts.

The cost of this factorization is that additional fields, the displacement field D and the magnetizing field H, are defined and need to be determined. Phenomenological constituent equations relate the additional fields to the electric field E and the magnetic B-field, often through a simple linear relation.

Bound Charge and Current

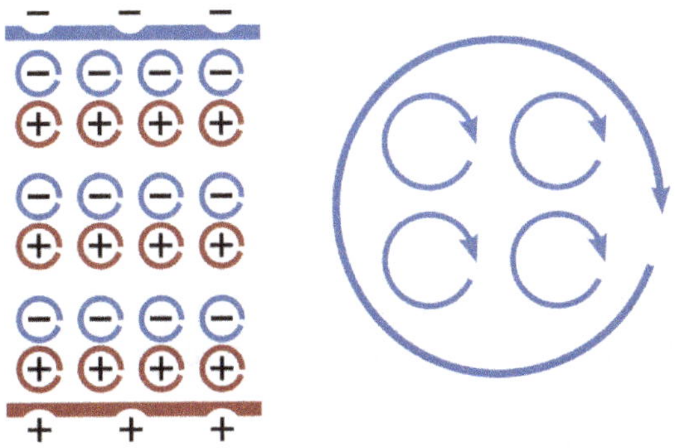

Left: A schematic view of how an assembly of microscopic dipoles produces opposite surface charges as shown at top and bottom. *Right:* How an assembly of microscopic current loops add together to produce a macroscopically circulating current loop. Inside the boundaries, the individual contributions tend to cancel, but at the boundaries no cancelation occurs.

When an electric field is applied to a dielectric material its molecules respond by forming microscopic electric dipoles – their atomic nuclei move a tiny distance in the direction of the field, while their electrons move a tiny distance in the opposite direction. This produces a *macroscopic bound charge* in the material even though all of the charges involved are bound to individual molecules. For example, if every molecule responds the same, similar to that shown in the figure, these tiny movements of charge combine to produce a layer of positive bound charge on one side of the material and a layer of negative charge on the other side. The bound charge is most conveniently described in terms of the polarization P of the material, its dipole moment per unit volume. If P is uniform, a macroscopic separation of charge is produced only at the surfaces where P enters and leaves the material. For non-uniform P, a charge is also produced in the bulk.

Somewhat similarly, in all materials the constituent atoms exhibit magnetic moments that are intrinsically linked to the angular momentum of the components of the atoms, most notably their electrons. The connection to angular momentum suggests the picture of an assembly of microscopic current loops. Outside the material, an assembly of such microscopic current loops is not different from a macroscopic current circulating around the material's surface, despite the fact that no individual charge is traveling a large distance. These *bound currents* can be described using the magnetization M.

The very complicated and granular bound charges and bound currents, therefore, can be represented on the macroscopic scale in terms of P and M which average these charges and currents on a sufficiently large scale so as not to see the granularity of individual atoms, but also sufficiently small that they vary with location in the material. As such, *Maxwell's macroscopic equations* ignore many details on a fine scale that can be unimportant to understanding matters on a gross scale by calculating fields that are averaged over some suitable volume.

Auxiliary Fields, Polarization and Magnetization

The *definitions* (not constitutive relations) of the auxiliary fields are:

$$\begin{aligned}\mathbf{D}(\mathbf{r},t) &= \varepsilon_0 \mathbf{E}(\mathbf{r},t) + \mathbf{P}(\mathbf{r},t)\\ \mathbf{H}(\mathbf{r},t) &= \frac{1}{\mu_0}\mathbf{B}(\mathbf{r},t) - \mathbf{M}(\mathbf{r},t)\end{aligned}$$

where P is the polarization field and M is the magnetization field which are defined in terms of microscopic bound charges and bound currents respectively. The macroscopic bound charge density ρ_b and bound current density J_b in terms of polarization P and magnetization M are then defined as

$$\begin{aligned}\rho_b &= -\nabla.\mathrm{P}\\ \mathrm{J}_b &= \nabla \times \mathrm{M} + \frac{\partial \mathrm{P}}{\partial \mathrm{t}}\end{aligned}$$

If we define the total, bound, and free charge and current density by

$$\begin{aligned}\rho &= \rho_b + \rho_f,\\ \mathbf{J} &= \mathbf{J}_b + \mathbf{J}_f,\end{aligned}$$

and use the defining relations above to eliminate D, and H, the "macroscopic" Maxwell's equations reproduce the "microscopic" equations.

Constitutive Relations

In order to apply 'Maxwell's macroscopic equations', it is necessary to specify the relations between displacement field D and the electric field E, as well as the magnetizing field H and the magnetic field B. Equivalently, we have to specify the dependence of the polarisation P (hence the bound charge) and the magnetisation M (hence the bound current) on the applied electric and magnetic field. The equations specifying this response are called constitutive relations. For real-world materials, the constitutive relations are rarely simple, except approximately, and usually determined by experiment.

For materials without polarisation and magnetisation, the constitutive relations are (by definition)

$$\mathbf{D} = \varepsilon_0 \mathbf{E}, \quad \mathbf{H} = \frac{1}{\mu_0}\mathbf{B}$$

where ε_0 is the permittivity of free space and μ_0 the permeability of free space. Since there is no bound charge, the total and the free charge and current are equal.

An alternative viewpoint on the microscopic equations is that they are the macroscopic equations *together* with the statement that vacuum behaves like a perfect linear "material" without additional polarisation and magnetisation. More generally, for linear materials the constitutive relations are

$$\mathrm{D} = \varepsilon \mathrm{E}, \quad \mathrm{H} = \frac{1}{\mu}\mathrm{B}$$

where ε is the permittivity and μ the permeability of the material. For the displacement field D the linear approximation is usually excellent because for all but the most extreme electric fields or temperatures obtainable in the laboratory (high power pulsed lasers) the interatomic electric fields of materials of the order of 10^{11} V/m are much higher than the external field. For the magnetizing field H, however, the linear approximation can break down in common materials like iron leading to phenomena like hysteresis. Even the linear case can have various complications, however.

- For homogeneous materials, ε and μ are constant throughout the material, while for inhomogeneous materials they depend on location within the material (and perhaps time).
- For isotropic materials, ε and μ are scalars, while for anisotropic materials (e.g. due to crystal structure) they are tensors.
- Materials are generally dispersive, so ε and μ depend on the frequency of any incident EM waves.

Even more generally, in the case of non-linear materials, D and P are not necessarily proportional to E, similarly H or M is not necessarily proportional to B. In general D and H depend on both E and B, on location and time, and possibly other physical quantities.

In applications one also has to describe how the free currents and charge density behave in terms of E and B possibly coupled to other physical quantities like pressure, and the mass, number density, and velocity of charge-carrying particles. E.g., the original equations given by Maxwell included Ohms law in the form

$$\mathbf{J}_{\mathrm{f}} = \sigma \mathbf{E}.$$

Alternative Formulations

Following is a summary of some of the numerous other ways to write the microscopic Maxwell's equations, showing they can be formulated using different mathematical formalisms. In addition, we formulate the equations using "potentials". Originally they were introduced as a convenient way to solve the homogeneous equations, but it was originally thought that all the observable physics was contained in the electric and magnetic fields (or relativistically, the Faraday tensor). The potentials play a central role in quantum mechanics, however, and act quantum mechanically with observable consequences even when the electric and magnetic fields vanish (Aharonov–Bohm effect). SI units are used throughout.

Formalism	Formulation	Homogeneous equations	Inhomogeneous equations
	Fields **3D Euclidean space + time**	$\nabla \cdot \mathbf{B} = 0$ $\nabla \times \mathbf{E} + \dfrac{\partial \mathbf{B}}{\partial t} = 0$	$\nabla \cdot \mathrm{E} = \dfrac{\rho}{\varepsilon_0}$ $\nabla \times \mathrm{B} - \dfrac{1}{\mathrm{c}^2}\dfrac{\partial \mathrm{E}}{\partial t} = \mu_0 \mathrm{J}$
Vector calculus	**Potentials (any gauge)** **3D Euclidean space + time**	$\mathbf{B} = \nabla \times \mathbf{A}$ $\mathbf{E} = -\nabla \varphi - \dfrac{\partial \mathbf{A}}{\partial t}$	$-\nabla^2 \varphi - \dfrac{\partial}{\partial t}(\nabla \cdot \mathbf{A}) = \dfrac{\rho}{\varepsilon_0}$ $(-\nabla^2 + \dfrac{1}{\mathrm{c}^2}\dfrac{\partial^2}{\partial t^2})\mathbf{A} + \nabla\left(\nabla \cdot \mathbf{A} + \dfrac{1}{\mathrm{c}^2}\dfrac{\partial \varphi}{\partial t}\right) = \mu_0 \mathbf{J}$
	Potentials (Lorenz gauge) **3D Euclidean space + time**	$\mathbf{B} = \nabla \times \mathbf{A}$ $\mathbf{E} = -\nabla \varphi - \dfrac{\partial \mathbf{A}}{\partial t}$ $\nabla \cdot \mathbf{A} = -\dfrac{1}{c^2}\dfrac{\partial \varphi}{\partial t}$	$(-\nabla^2 + \dfrac{1}{\mathrm{c}^2}\dfrac{\partial^2}{\partial t^2})\varphi = \dfrac{\rho}{\varepsilon_0}$ $(-\nabla^2 + \dfrac{1}{\mathrm{c}^2}\dfrac{\partial^2}{\partial t^2})\mathbf{A} = \mu_0 \mathbf{J}$

Tensor calculus	**Fields** **space + time** **spatial metric independent of time**	$\partial_{[i}B_{jk]} =$ $\nabla_{[i}B_{jk]} = 0$ $\partial_{[i}E_{j]} + \frac{\partial B_{ij}}{\partial t} =$ $\nabla_{[i}E_{j]} + \frac{\partial B_{ij}}{\partial t} = 0$	$\frac{1}{\sqrt{h}}\partial_i \sqrt{h}E^i =$ $\nabla_i E^i = \frac{\rho}{\epsilon_0}$ $-\frac{1}{\sqrt{h}}\partial_i \sqrt{h}B^{ij} - \frac{1}{c^2}\frac{\partial}{\partial t}E^j =$ $-\nabla_i B^{ij} - \frac{1}{c^2}\frac{\partial E^j}{\partial t} = \mu_0 J^j$
	Potentials **space (with topological restrictions) + time** **spatial metric independent of time**	$B_{ij} = \partial_{[i}A_{j]}$ $= \nabla_{[i}A_{j]}$ $E_i = -\frac{\partial A_i}{\partial t} - \partial_i \phi$ $= -\frac{\partial A_i}{\partial t} - \nabla_i \phi$	$-\frac{1}{\sqrt{h}}\partial_i \sqrt{h}(\partial^i \phi + \frac{\partial A^i}{\partial t}) =$ $-\nabla_i \nabla^i \phi - \frac{\partial}{\partial t}\nabla_i A^i = \frac{\rho}{\epsilon_0}$ $-\frac{1}{\sqrt{h}}\partial_i(\sqrt{h}h^{im}h^{jn}\partial_{[m}A_{n]}) + \frac{1}{c^2}\frac{\partial}{\partial t}(\frac{\partial A^j}{\partial t} + \partial^j \phi) =$ $-\nabla_i \nabla^i A^j + \frac{1}{c^2}\frac{\partial^2 A^j}{\partial t^2} + R_i^j A^i + \nabla^j(\nabla_i A^i + \frac{1}{c^2}\frac{\partial \phi}{\partial t}) = \mu_0 J^j$
	Potentials (Lorenz gauge) **space (with topological restrictions) + time** **spatial metric independent of time**	$B_{ij} = \partial_{[i}A_{j]}$ $= \nabla_{[i}A_{j]}$ $E_i = -\frac{\partial A_i}{\partial t} - \partial_i \phi$ $= -\frac{\partial A_i}{\partial t} - \nabla_i \phi$ $\nabla_i A^i = -\frac{1}{c^2}\frac{\partial \phi}{\partial t}$	$-\nabla_i \nabla^i \phi + \frac{1}{c^2}\frac{\partial^2 \phi}{\partial t^2} = \frac{\rho}{\epsilon_0}$ $-\nabla_i \nabla^i A^j + \frac{1}{c^2}\frac{\partial^2 A^j}{\partial t^2} + R_i^j A^i = \mu_0 J^j$
Differential forms	**Fields** **Any space + time**	$dB = 0$ $dE + \frac{\partial B}{\partial t} = 0$	$d{*}E = \frac{\rho}{\epsilon_0}$ $d{*}B - \frac{1}{c^2}\frac{\partial {*}E}{\partial t} = \mu_0 J$
	Potentials (any gauge) **Any space (with topological restrictions) + time**	$B = dA$ $E = -d\phi - \frac{\partial A}{\partial t}$	$-d{*}(d\phi + \frac{\partial A}{\partial t}) = \frac{\rho}{\epsilon_0}$ $d{*}dA + \frac{1}{c^2}\frac{\partial}{\partial t}{*}(d\phi - \frac{\partial A}{\partial t}) = \mu_0 J$
	Potential (Lorenz Gauge) **Any space (with topological restrictions) + time** **spatial metric independent of time**	$B = dA$ $E = -d\phi - \frac{\partial A}{\partial t}$ $d{*}A = -{*}\frac{1}{c^2}\frac{\partial \phi}{\partial t}$	${*}(-\Delta\phi + \frac{1}{c^2}\frac{\partial^2}{\partial t^2}\phi) = \frac{\rho}{\epsilon_0}$ ${*}(-\Delta A + \frac{1}{c^2}\frac{\partial^2 A}{\partial^2 t}) = \mu_0 J$

where

- In the vector formulation on Euclidean space + time, φ is the electrical potential, and A is the vector potential.

Relativistic Formulations

The Maxwell equations can also be formulated on a space-time like Minkowski space where space and time are treated on equal footing. The direct space–time formulations make manifest that the Maxwell equations are relativistically invariant. Because of this symmetry electric and magnetic field are treated on equal footing and are recognised as components of the Faraday tensor. This reduces the four Maxwell equations to two, which simplifies the equations, although we can no longer use the familiar vector formulation. In fact the Maxwell equations in the space + time formulation are not Galileo invariant and have Lorenz invariance a hidden symmetry. This was a major source of inspiration for the development of relativity theory. The space + time formulation is not a non-relativistic approximation, however, they describe the same physics by simply renaming variables. For this reason the relativistic invariant equations are usually simply called the Maxwell equations as well.

Formalism	Formulation	Homogeneous equations	Inhomogeneous equations
Tensor calculus	**Fields** **Minkowski space**	$\partial_{[\alpha}F_{\beta\gamma]}=0$	$\partial_\alpha F^{\alpha\beta}=\mu_0 J^\beta$
	Potentials (any gauge) **Minkowski space**	$F_{\alpha\beta}=2\partial_{[\alpha}A_{\beta]}$	$2\partial_\alpha\partial^{[\alpha}A^{\beta]}=\mu_0 J^\beta$
	Potentials (Lorenz gauge) **Minkowski space**	$F_{\alpha\beta}=2\partial_{[\alpha}A_{\beta]}$ $\partial_\alpha A^\alpha=0$	$\partial_\alpha\partial^\alpha A^\beta=\mu_0 J^\beta$
	Fields **Any space–time**	$\partial_{[\alpha}F_{\beta\gamma]}=\nabla_{[\alpha}F_{\beta\gamma]}=0$	$\frac{1}{\sqrt{-g}}\partial_\alpha(\sqrt{-g}F^{\alpha\beta})=\nabla_\alpha F^{\alpha\beta}=\mu_0 J^\beta$
	Potentials (any gauge) **Any space–time (with topological restrictions)**	$F_{\alpha\beta}=2\partial_{[\alpha}A_{\beta]}=2\nabla_{[\alpha}A_{\beta]}$	$\frac{2}{\sqrt{-g}}\partial_\alpha(\sqrt{-g}g^{\alpha\mu}g^{\beta\nu}\partial_{[\mu}A_{\nu]})=2\nabla_\alpha(\nabla^{[\alpha}A^{\beta]})=\mu_0 J^\beta$
	Potentials (Lorenz gauge) **Any space–time (with topological restrictions)**	$F_{\alpha\beta}=2\partial_{[\alpha}A_{\beta]}=2\nabla_{[\alpha}A_{\beta]}$ $\nabla_\alpha A^\alpha=0$	$\nabla_\alpha\nabla^\alpha A^\beta-R^\beta{}_\alpha A^\alpha=\mu_0 J^\beta$

Differential forms			
	Fields **Any space–time**	$\mathrm{dF} = 0$	$\mathrm{d} \star F = \mu_0 J$
	Potentials (any gauge) **Any space–time (with topological restrictions)**	$\mathrm{F} = \mathrm{dA}$	$\mathrm{d} \star \mathrm{d}A = \mu_0 J$
	Potentials (Lorenz gauge) **Any space–time (with topological restrictions)**	$F = \mathrm{d}A$ $\mathrm{d} \star A = 0$	$\star \Box \mathrm{A} = \mu_0 \mathrm{J}$

- In the tensor calculus formulation, the electromagnetic tensor $F_{\alpha\beta}$ is an antisymmetric covariant rank 2 tensor; the four-potential, A_α, is a covariant vector; the current, J^α, is a vector; the square brackets, [ ], denote antisymmetrization of indices; ∂_α is the derivative with respect to the coordinate, x^α. In Minkowski space coordinates are chosen with respect to an inertial frame; $(x^\alpha) = (ct,x,y,z)$, so that the metric tensor used to raise and lower indices is $\eta_{\alpha\beta} = \mathrm{diag}(1,-1,-1,-1)$. The d'Alembert operator on Minkowski space is $\Box = \partial_\alpha \partial^\alpha$ as in the vector formulation. In general spacetimes, the coordinate system x^α is arbitrary, the covariant derivative ∇_α, the Ricci tensor, $R_{\alpha\beta}$ and raising and lowering of indices are defined by the Lorentzian metric, $g_{\alpha\beta}$ and the d'Alembert operator is defined as $\Box = \nabla_\alpha \nabla^\alpha$. The topological restriction is that the second real cohomology group of the space vanishes. Note that this is violated for Minkowski space with a line removed, which can model a (flat) space-time with a point-like monopole on the complement of the line.

- In the differential form formulation on arbitrary space times, $F = F_{\alpha\beta} \mathrm{d}x^\alpha \wedge \mathrm{d}x^\beta$ is the electromagnetic tensor considered as a 2-form, $A = A_\alpha \mathrm{d}x^\alpha$ is the potential 1-form, J is the current 3-form, d is the exterior derivative, and $\star$ is the Hodge star $\star$ on forms defined by the Lorentzian metric of space–time. Note that in the special case of 2-forms such as F, the Hodge star only depends on the metric up to a local scale. This means that, as formulated, the differential form field equations are conformally invariant, but the Lorenz gauge condition breaks conformal invariance. The operator $\Box = (-\star \mathrm{d} \star \mathrm{d} - \mathrm{d} \star \mathrm{d} \star)$ is the d'Alembert–Laplace–Beltrami operator on 1-forms on an arbitrary Lorentzian space–time. The topological condition is again that the second real cohomology group is trivial. By the isomorphism with the second de Rham cohomology this condition means that every closed 2 form is exact.

Other formalisms include the geometric algebra formulation and a matrix representation of Maxwell's equations. Historically, a quaternionic formulation was used.

Solutions

Maxwell's equations are partial differential equations that relate the electric and magnetic fields to each other and to the electric charges and currents. Often, the charges and currents are themselves dependent on the electric and magnetic fields via the Lorentz force equation and the constitutive relations. These all form a set of coupled partial differential equations, which are often very difficult to solve. In fact, the solutions of these equations encompass all the diverse phenomena in the entire field of classical electromagnetism.

Like any differential equation, boundary conditions and initial conditions are necessary for a unique solution. For example, even with no charges and no currents anywhere in spacetime, many solutions to Maxwell's equations are possible, not just the obvious solution E = B = 0. Another solution is E = constant, B = constant, while yet other solutions have electromagnetic waves filling spacetime. In some cases, Maxwell's equations are solved through infinite space, and boundary conditions are given as asymptotic limits at infinity. In other cases, Maxwell's equations are solved in just a finite region of space, with appropriate boundary conditions on that region: For example, the boundary could be an artificial absorbing boundary representing the rest of the universe, or periodic boundary conditions, or (as with a waveguide or cavity resonator) the boundary conditions may describe the walls that isolate a small region from the outside world.

Jefimenko's equations (or the closely related Liénard–Wiechert potentials) are the explicit solution to Maxwell's equations for the electric and magnetic fields created by any given distribution of charges and currents. It assumes specific initial conditions to obtain the so-called "retarded solution", where the only fields present are the ones created by the charges. Jefimenko's equations are not so helpful in situations when the charges and currents are themselves affected by the fields they create.

Numerical methods for differential equations can be used to approximately solve Maxwell's equations when an exact solution is impossible. These methods usually require a computer, and include the finite element method and finite-difference time-domain method.

Maxwell's equations *seem* overdetermined, in that they involve six unknowns (the three components of E and B) but eight equations (one for each of the two Gauss's laws, three vector components each for Faraday's and Ampere's laws). (The currents and charges are not unknowns, being freely specifiable subject to charge conservation.) This is related to a certain limited kind of redundancy in Maxwell's equations: It can be proven that any system satisfying Faraday's law and Ampere's law *automatically* also satisfies the two Gauss's laws, as long as the system's initial condition does. This explanation was first introduced by Julius Adams Stratton in 1941. Although it is possible to simply ignore the two Gauss's laws in a numerical algorithm (apart from the initial conditions), the imperfect precision of the calculations can lead to ever-increasing violations of those laws. By introducing dummy variables characterizing these violations, the four equations become not overdetermined after all. The resulting formulation can lead to more accurate algorithms that take all four laws into account.

Limitations of the Maxwell Equations as a Theory of Electromagnetism

While Maxwell's equations (along with the rest of classical electromagnetism) are extraordinarily successful at explaining and predicting a variety of phenomena, they are not exact, but approximations. In some special situations, they can be noticeably inaccurate. Examples include extremely strong fields and extremely short distances. Moreover, various phenomena occur in the world even though Maxwell's equations predict them to be impossible, such as "nonclassical light" and quantum entanglement of electromagnetic fields. Finally, any phenomenon involving individual photons, such as the photoelectric effect, Planck's law, the Duane–Hunt law, single-photon light detectors, etc., would be difficult or impossible to explain if Maxwell's equations were exactly true, as Maxwell's equations do not involve photons. For the most accurate predictions in all situations, Maxwell's equations have been superseded by quantum electrodynamics.

Variations

Popular variations on the Maxwell equations as a classical theory of electromagnetic fields are relatively scarce because the standard equations have stood the test of time remarkably well.

Magnetic Monopoles

Maxwell's equations posit that there is electric charge, but no magnetic charge (also called magnetic monopoles), in the universe. Indeed, magnetic charge has never been observed (despite extensive searches) and may not exist. If they did exist, both Gauss's law for magnetism and Faraday's law would need to be modified, and the resulting four equations would be fully symmetric under the interchange of electric and magnetic fields.

Maxwell's Equations in Vacuum

In the Language of Differential Vector Calculus

- Gauss's law: $\nabla.\vec{E} = \frac{\rho}{\varepsilon_0}$
- Gauss's law for magnetism: $\nabla.\vec{B} = 0$
- Maxwell-Faraday equation: $\nabla \times \vec{E} = -\frac{\partial \vec{B}}{\partial t}$

- Ampere's law with Maxwell's correction: $\nabla \times \vec{B} = \mu_0 \left(\vec{J} + \varepsilon_0 \frac{\partial \vec{E}}{\partial t} \right)$

We shall now look at interpretations of these expressions by using their integral forms.

Gauss's Law: Enclosed Charges

$$\nabla . \vec{E} = \rho / \varepsilon_0 :$$

- Integrate over a closed volume:

$$\int_v (\nabla . \vec{E})\, dV = \int_v \frac{\rho}{\varepsilon_0} dV$$

- Use a mathematical identity (Gauss's theorem)

$$\oint \vec{E} . d\vec{S} = \frac{Q_{enclosed}}{\varepsilon_0}$$

- Relationship between electric field on a closed surface and the charge enclosed inside it
- The part in red: source of the electric field
- Leads to Coulomb's law if Q is a point charge at the centre of $\vec{S}$, a sphere of radius r :

$$E_r . 4\pi r^2 = Q / \varepsilon_0$$

Gauss's Law: no Magnetic Monopoles

$$\nabla . \vec{B} = 0 :$$

- Integrate over a closed volume:

$$\int_v (\nabla . \vec{B})\, dV = 0$$

- Use a mathematical identity (Gauss's theorem)

$$\oint \vec{B} . d\vec{S} = 0$$

- Relationship between magnetic field on a closed surface and the charge enclosed inside it

- The part in red: source of the magnetic field
- Vanishing of the source ⇒ no magnetic monopoles

Maxwell-Faraday Equation: Flux Through a Loop

$$\nabla \times \vec{E} = -\partial \vec{B} / \partial t:$$

- Integrate over a surface whose boundary is a loop:

$$\int_{\vec{S}} (\nabla \times \vec{E}) \,.\, d\vec{S} = \int_{\vec{S}} -(\partial \vec{B} / \partial t) \,.\, d\vec{S}$$

- Use a mathematical identity (Stokes' theorem)

$$\oint \vec{E} \,.\, d\vec{\ell} = -\int_{\vec{S}} \frac{\partial}{\partial t} (\vec{B} \,.\, d\vec{S})$$

(If the loop does not change with time)

- The induced EMF is

$$\varepsilon \equiv \oint \vec{E} \,.\, d\vec{\ell} = -\frac{\partial}{\partial t} \int_{\vec{S}} (\vec{B} . d\vec{S}) = -\frac{\partial \phi}{\partial t}$$

$$\phi \equiv \int_{\vec{S}} \vec{B} \,.\, d\vec{S}$$

More Comments on the Maxwell-Faraday Equation

- Relationship between electric field along a loop and the rate of change of magnetic flux through an open surface whose boundary is the loop
- No sources needed: it is a relationship between $\vec{E}$ and $\vec{B}$
- The "$\varepsilon = -\partial \phi / \partial t$" equation does not hold for all situations, since it does not take into account the Lorentz force on a moving charge in a magnetic field.

Ampere's Law with Maxwell's Corrections

$$\nabla \times \vec{B} = \mu_0 (\vec{J} + \varepsilon_0 \partial \vec{E} / \partial t):$$

- Integrate over a surface whose boundary is a loop:

$$\int_{\vec{S}} (\nabla \times \vec{B}) \,.\, d\vec{S} = \mu_0 \int_{\vec{S}} \vec{J} . \, d\vec{S} + \mu_0 \varepsilon_0 \int_{\vec{S}} (\partial \vec{E} / \partial t) \,.\, d\vec{S}$$

- Use a mathematical identity (Stokes' theorem)

$$\oint \vec{B}.\,d\vec{\ell} = \mu_0 I + \mu_0\varepsilon_0 \int_{\bar{S}} \frac{\partial}{\partial t}(\vec{E}.\,d\vec{S})$$

- Relationship between magnetic field along a loop and the rate of change of magnetic flux through an open surface whose boundary is the loop
- $I = \oint_{\bar{S}} \vec{J}.\,d\vec{S}$ is the conduction current
- $\mu_0 \int_{\bar{S}} \frac{\partial}{\partial t}(\vec{E}.d\vec{S})$ is often called "displacement current", this is the correction by Maxwell to Ampere's law

Inside a Dielectric Medium (Static Case)

- Gauss's law always valid, when ρ is the total charge: $\nabla.\vec{E} = \rho / \varepsilon_0$
- Part of the charge is due to polarization induced in the medium, which gives rise to the "bound charge": $\rho b = -\nabla.\vec{p}$ where $\vec{p}$ is the polarization
- Then $\varepsilon_0 \nabla.\vec{E} = (\rho b + \rho_{fr}) = -\nabla.P + \rho_{fr}$

  where ρ_{fr} is the free charge density
- Defining $\vec{D} = \varepsilon_0 \vec{E} + \vec{P}$, we get Gauss's law in terms of the free charge density:

$$\nabla.\vec{D} = \rho_{fr}$$

- The relation $\vec{D} = \varepsilon \vec{E}$ defines the dielectric permittivity of the medium, ε. This is in general not a number but a tensor, and may not be constant. Wherever it is constant, the dielectric is called "linear".

Inside a Magnetic Medium (Static Case)

- Maxwell-Faraday equation always valid, when $\vec{J}$ is the total current: $\nabla \times \vec{B} = \mu_0 \vec{J}$
- Part of the current is due to magnetization induced in the medium, which gives rise to the "surface current": $\vec{J}_{surface} = \nabla \times \vec{\mathbf{M}}$, where $\vec{\mathbf{M}}$ is the magnetization
- Then $\nabla \times \vec{B} = (\vec{J}_{surface} + \vec{J}_{fr}) = \mu_0 \nabla \times M + \mu_0 \vec{J}_{fr}$

  where $\vec{J}_{fr}$ is the free current density
- Defining $\vec{H} = \vec{B} / \mu_0 - \vec{M}$, we get Ampere's law in terms of the free charge density:

$$\nabla \times \vec{H} = \vec{J}_{fr}$$

- The relation $\vec{B} = \mu \vec{H}$ defines the magnetic permeability of the medium, μ. This is in general not a number but a tensor, and may not be constant. Wherever it is constant, the magnetic medium is called “linear”.

Maxwell’s Equations: “Macroscopic” Form

$$\nabla.\vec{D} = \rho_{fr}$$

$$\nabla.\vec{B} = 0$$

$$\nabla \times \vec{E} = -\frac{\partial \vec{B}}{\partial t}$$

$$\nabla \times \vec{B} = \vec{J}_{fr} + \frac{\partial \vec{D}}{\partial t}$$

These are equivalent to the equations (1)–(4), with the substitutions

$$\rho = \rho_{fr} + \rho_b , \quad \vec{J} = \vec{J}_{fr} + \vec{J}_{surface}$$

$$\vec{D} = \varepsilon_0 \vec{E} + \vec{P} , \quad \vec{B} = \mu_0 (\vec{H} + \vec{M})$$

$$\rho_b = -\nabla.\vec{P} , \vec{J}_{surface} = \nabla \times \vec{M} + \frac{\partial \vec{D}}{\partial t}.$$

Near and Far Field

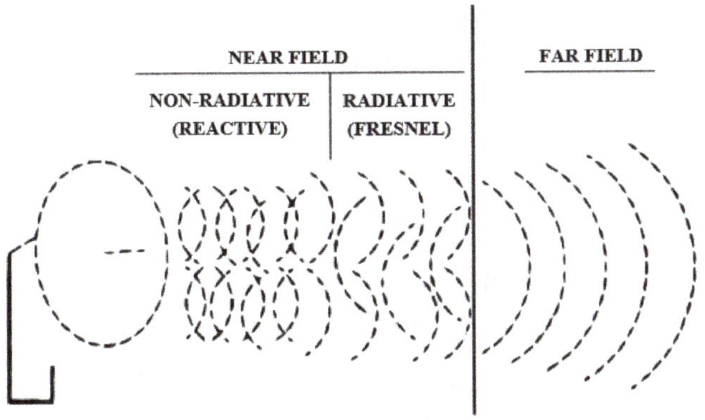

Differences between Fraunhofer diffraction and Fresnel diffraction

The near field and far field are regions of the electromagnetic field around an object, such as a transmitting antenna, or the result of radiation scattering off an object. Non-radiative ‘near-field’ behaviors of electromagnetic fields dominate close to the antenna or scattering object, while electromagnetic radiation ‘far-field’ behaviors dominate at greater distances.

Far-field E and B field strength decreases inversely with distance from the source, resulting in an inverse-square law for the radiated power intensity of electromagnetic radiation. By contrast, near-field E and B strength decrease more rapidly with distance (with inverse-distance squared or cubed), resulting in relative lack of near-field effects within a few wavelengths of the radiator.

Summary of Regions and their Interactions

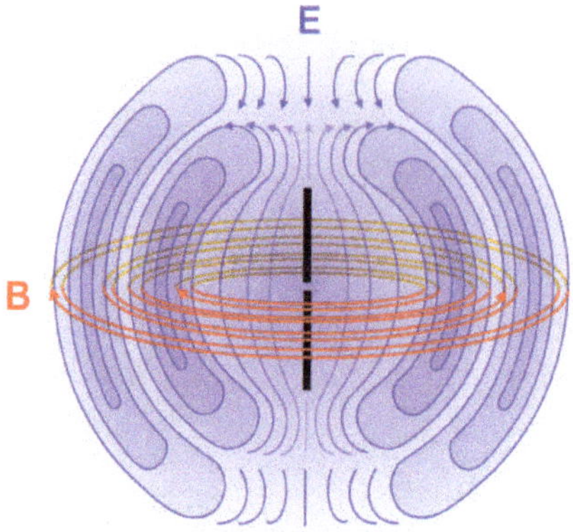

Near field: This dipole pattern shows a magnetic field $\vec{\mathbf{B}}$ in red. The potential energy momentarily stored in this magnetic field is indicative of the reactive near field.

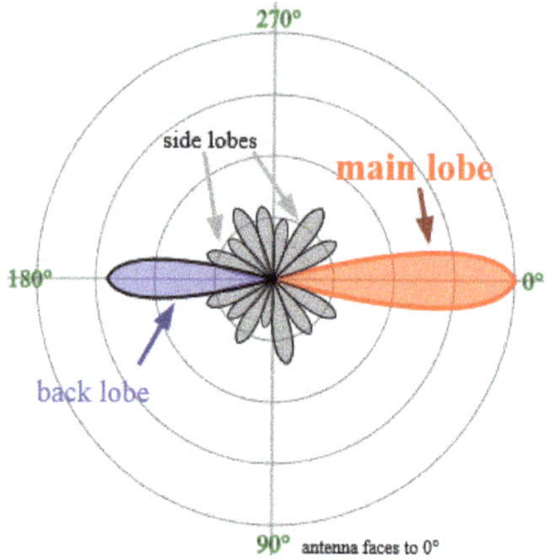

Far field: The radiation pattern can extend into the far field, where the reactive stored energy has no significant presence.

The far field is the region in which the field acts as "normal" electromagnetic radiation. In this region, it is dominated by electric or magnetic fields with electric dipole characteristics. The near field is governed by multipole type fields, which can be considered as collections of dipoles with a fixed phase relationship. The boundary between the two regions is only vaguely defined, and it depends on the dominant wavelength (λ) emitted by the source.

In the far-field region of an antenna, radiation decreases as the square of distance, and absorption of the radiation does not feed back to the transmitter. However, in the near-field region, absorption of radiation does affect the load on the transmitter. Magnetic induction (for example, in a transformer) can be seen as a very simple model of this type of near-field electromagnetic interaction.

In the far-field region, each part of the EM field (electric and magnetic) is "produced by" (or associated with) a change in the other part, and the ratio of electric and magnetic field intensities is simply the wave impedance. However, in the near-field region, the electric and magnetic fields can exist independently of each other, and one type of field can dominate the other.

In a normally-operating antenna, positive and negative charges have no way of leaving and are separated from each other by the excitation "signal" (a transmitter or other EM exciting potential). This generates an oscillating (or reversing) electrical dipole, which affects both the near field and the far field. In general, the purpose of antennas is to communicate wirelessly for long distances using far fields, and this is their main region of operation (however, certain antennas specialized for near-field communication do exist).

Also known as the radiation-zone field, the far field carries a relatively uniform wave pattern. The radiation zone is important because far fields in general fall off in amplitude by $1/r$. This means that the total energy per unit area at a distance r is proportional to $1/r^2$. The area of the sphere is proportional to r^2, so the total energy passing through the sphere is constant. This means that the far-field energy actually escapes to infinite distance (it *radiates*).

In contrast, the near field refers to regions such as near conductors and inside polarizable media where the propagation of electromagnetic waves is interfered with. One easy to observe example is the change of noise levels picked up by a set of rabbit ear antennas when one places a body part in close range. The near-field has been of increasing interest, particularly in the development of capacitive sensing technologies such as those used in smart phones and tablet computers.

The interaction with the medium (e.g. body capacitance) can cause energy to deflect back to the source, as occurs in the *reactive* near field. Or the interaction with the medium can fail to return energy back to the source, but cause a distortion in the electromagnetic wave that deviates significantly from that found in a hard vacuum, and this indicates the *radiative* near-field region, which is somewhat further away. Another intermediate region, called the *transition zone*, is defined on a somewhat different basis, namely antenna geometry and excitation wavelength.

Definitions

The term "near-field region" (also known as the "near field" or "near zone") has the following meanings with respect to different telecommunications technologies:

- The close-in region of an antenna where the angular field distribution is dependent upon the distance from the antenna.

- In the study of diffraction and antenna design, the near field is that part of the radiated field that is below distances shorter than the Fraunhofer distance $d_f = 2D^2/\lambda$ from the source of the diffracting edge or antenna of longitude or diameter D.
- In optical fiber communications, the region near a source or aperture that is closer than the Rayleigh length. (Presuming a Gaussian beam, which is appropriate for fiber optics)

Because of these nuances, special care must be taken when comprehending the literature about near fields and far fields.

Regions According to Electromagnetic Length

Electromagnetically Short Antennas

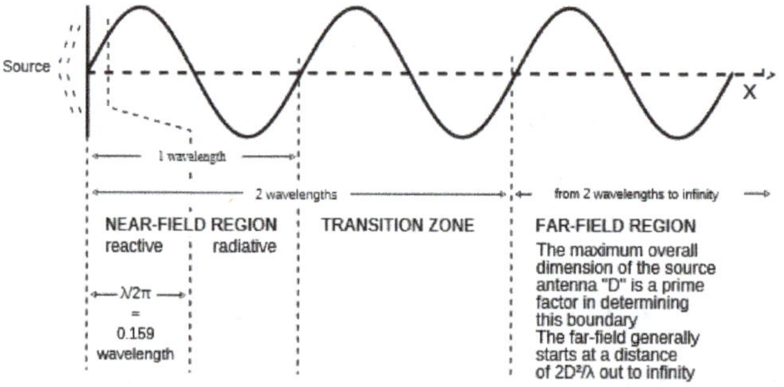

Field regions for antennas equal to, or shorter than, one-half wavelength of the radiation they emit, such as the whip antenna of a citizen's band radio, or an AM radio broadcast tower.

For antennas *shorter than half of the wavelength of the radiation they emit* (i.e., "electromagnetically short" antennas), the far and near regional boundaries are measured in terms of a simple ratio of the distance r from the radiating source to the wavelength λ of the radiation. For such an antenna, the near field is the region within a radius ($r \ll \lambda$), while the far-field is the region for which $r \gg 2\lambda$. The transition zone is the region between $r = \lambda$ and $r = 2\lambda$.

Note that D, the length of the antenna is not important, and the approximation is the same for all shorter antennas (sometimes ideally called "point antennas"). In all such antennas, the short length means that charges and currents in each sub-section of the antenna are the same at any given time, since the antenna is too short for the RF transmitter voltage to reverse before its effects on charges and currents are felt over the entire antenna length.

Electromagnetically Long Antennas

For antennas physically larger than a half-wavelength of the radiation they emit, the near and far fields are defined in terms of the Fraunhofer distance. The Fraunhofer distance, named after Joseph von Fraunhofer, is given by the following:

$$d_f = \frac{2D^2}{\lambda},$$

where *D* is the largest dimension of the radiator (or the diameter of the antenna) and λ is the wavelength of the radio wave. Either of the following two relations are equivalent, emphasizing the size of the region in terms of wavelengths λ or diameters D:

$$d_f = 2\left(\frac{D}{\lambda}\right)^2 \lambda = 2\left(\frac{D}{\lambda}\right) D,$$

This distance provides the limit between the near and far field. The parameter *D* corresponds to the physical length of an antenna, or the diameter of a "dish" antenna.

Having an antenna electromagnetically longer than one-half the dominated wavelength emitted considerably extends the near-field effects, especially that of focused antennas. Conversely, when a given antenna emits high frequency radiation, it will have a near-field region larger than what would be implied by the shorter wavelength.

Additionally, a far-field region distance d_f must satisfy these two conditions.

$$d_f \gg D,$$
$$d_f \gg \lambda,$$

where *D* is the largest physical linear dimension of the antenna and d_f is the far-field distance. The far-field distance is the distance from the transmitting antenna to the beginning of the Fraunhofer region, or far field.

Transition Zone

The "transition zone" between these near and far field regions, extending over the distance from one to two wavelengths from the antenna, is the intermediate region in which both near-field and far-field effects are important. In this region, near-field behavior dies out and ceases to be important, leaving far-field effects as dominant interactions.

Regions According to Diffraction Behavior

Far-field Diffraction

As far as acoustic wave sources are concerned, if the source has a maximum overall dimension or aperture width (*D*) that is large compared to the wavelength λ, the far-field region is commonly taken to exist at distances from the source greater than Fresnel parameter $S = D^2/(4\lambda)$, $S > 1$.

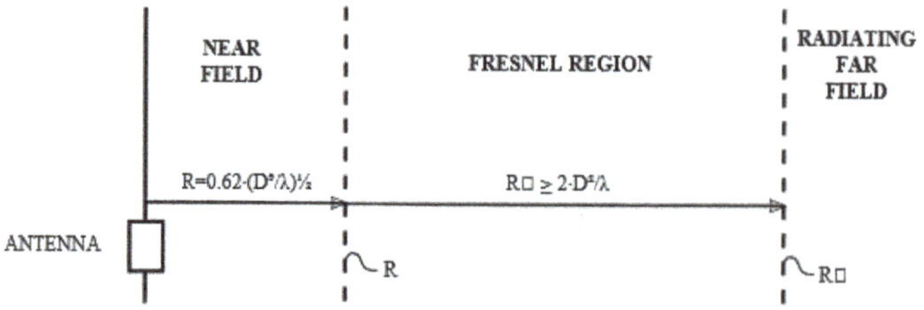

Near- and far-field regions for an antenna larger (diameter or length D) than the wavelength of the radiation it emits, so that $D/\lambda \gg 1$. Examples are radar dishes and other highly directional antennas.

For a beam focused at infinity, the far-field region is sometimes referred to as the "Fraunhofer region". Other synonyms are "far field", "far zone", and "radiation field". Any electromagnetic radiation consists of an electric field component E and a magnetic field component H. In the far field, the relationship between the electric field component E and the magnetic component H is that characteristic of any freely propagating wave, where (in units where $c = 1$) E and H have equal magnitudes at any point in space.

Near-field Diffraction

In contrast to the far field, the diffraction pattern in the near field typically differs significantly from that observed at infinity and varies with distance from the source. In the near field, the relationship between E and H becomes very complex. Also, unlike the far field where electromagnetic waves are usually characterized by a single polarization type (horizontal, vertical, circular, or elliptical), all four polarization types can be present in the near field.

The "near field" is a region in which there are strong inductive and capacitive effects from the currents and charges in the antenna that cause electromagnetic components that do not behave like far-field radiation. These effects decrease in power far more quickly with distance than do the far-field radiation effects. Non-propagating (or evanescent) fields extinguish very rapidly with distance, which makes their effects almost exclusively felt in the near-field region.

Also, in the part of the near field closest to the antenna, absorption of electromagnetic power in the region by a second device has effects that feed back to the transmitter, increasing the load on the transmitter that feeds the antenna by decreasing the antenna impedance that the transmitter "sees". Thus, the transmitter can sense when power is being absorbed in the closest near-field zone (by a second antenna or some other object) and is forced to supply extra power to its antenna, and to draw extra power from its own power supply, whereas if no power is being absorbed there, the transmitter does not have to supply extra power.

Variations Within Regions

The above defined regions categorize field behaviors that *vary*, even within the region of interest. Thus, the boundaries for these regions are approximate "rules of thumb", as there are no precise cutoffs between them (all behavioral changes with distance are smooth changes). Even when precise boundaries can be defined in some cases, based primarily on antenna type and antenna size, experts may differ in their use of nomenclature to describe the regions.

Near-field Characteristics

The near field itself is further divided into the reactive near field and the radiative near field. The "reactive" and "radiative" near-field designations are also a function of wavelength (or distance). However, these boundary regions are a fraction of one wavelength within the near field. The outer boundary of the reactive near-field region is commonly considered to be a distance of $1/2\pi$ times the wavelength ($\lambda/2\pi$ or $0.159 \times \lambda$) from the antenna surface. The radiative near field (also called the "Fresnel region") covers the remainder of the near-field region, from $\lambda/2\pi$ out to the Fraunhofer distance.

Reactive Near Field, or the Nearest Part of the Near Field

In the reactive near field (very close to the antenna), the relationship between the strengths of the E and H fields is often too complex to predict. Either field component (E or H) may dominate at one point, and the opposite relationship dominate at a point only a short distance away. This makes finding the true power density in this region problematic. This is because to calculate power, not only E and H both have to be measured but the phase relationship between E and H as well as the angle between the two vectors must also be known in every point of space.

In this reactive region, not only is an electromagnetic wave being radiated outward into far space but there is a "reactive" component to the electromagnetic field, meaning that the nature of the field around the antenna is sensitive to, and reacts to, EM absorption in this region (this is not true for absorption far from the antenna, which has no effect on the transmitter or antenna near field).

Very close to the antenna, in the reactive region, energy of a certain amount, if not absorbed by a receiver, is held back and is stored very near the antenna surface. This energy is carried back and forth from the antenna to the reactive near field by electromagnetic radiation of the type that slowly changes electrostatic and magnetostatic effects. For example, current flowing in the antenna creates a purely magnetic component in the near field, which then collapses as the antenna current begins to reverse, causing transfer of the field's magnetic energy back to electrons in the antenna as the changing magnetic field causes a self-inductive effect on the antenna that generated it. This returns energy to the antenna in a regenerative way, so that it is not lost. A similar process happens as electric charge builds up in one section of the antenna under the pressure of the signal voltage, and causes a local electric field around that section of antenna, due to the antenna's self-capacitance. When the signal reverses so that charge is allowed to flow away from this region again, the built-up electric field assists in pushing electrons back in the new direction of their flow, as with the discharge of any unipolar capacitor. This again transfers energy back to the antenna current.

Because of this energy storage and return effect, if either of the inductive or electrostatic effects in the reactive near field transfer any field energy to electrons in a different (nearby) conductor, then this energy is lost to the primary antenna. When this happens, an extra drain is seen on the transmitter, resulting from the reactive near-field energy that is not returned. This effect shows up as a different impedance in the antenna, as seen by the transmitter.

The reactive component of the near field can give ambiguous or undetermined results when attempting measurements in this region. In other regions, the power density is inversely proportional to the square of the distance from the antenna. In the vicinity very close to the antenna, however, the energy level can rise dramatically with only a small decrease in distance toward the antenna. This energy can adversely affect both humans and measurement equipment because of the high powers involved.

Radiative Near Field (Fresnel Region), or Farthest Part of the Near Field

The radiative near field (sometimes called the Fresnel region) does not contain reactive field components from the source antenna, since it is so far from the antenna that back-coupling of the fields becomes out of phase with the antenna signal, and thus cannot efficiently store and replace inductive or capacitive energy from antenna currents or charges. The energy in the radiative near field is thus all radiant energy, although its mixture of magnetic and electric components are still different from the far field. Further out into the radiative near field (one half wavelength to 1 wavelength from the source), the E and H field relationship is more predictable, but the E to H relationship is still complex. However, since the radiative near field is still part of the near field, there is potential for unanticipated (or adverse) conditions.

For example, metal objects such as steel beams can act as antennas by inductively receiving and then "re-radiating" some of the energy in the radiative near field, forming a new radiating surface to consider. Depending on antenna characteristics and frequencies, such coupling may be far more efficient than simple antenna reception in the yet-more-distant far field, so far more power may be transferred to the secondary "antenna" in this region than would be the case with a more distant antenna. When a secondary radiating antenna surface is thus activated, it then creates its own near-field regions, but the same conditions apply to them.

Compared to the Far Field

The near field is remarkable for reproducing classical electromagnetic induction and electric charge effects on the EM field, which effects "die-out" with increasing distance from the antenna (with magnetic field strength proportional to the inverse-cube of the distance and electric field strength proportional to inverse-square of distance), far more rapidly than do the classical radiated EM far-field (E and B fields proportional simply to inverse-distance). Typically near-field effects are not important farther away than a few wavelengths of the antenna.

More-distant near-field effects also involve energy transfer effects that couple directly to receivers near the antenna, affecting the power output of the transmitter if they do couple, but not otherwise. In a sense, the near field offers energy that is available to a receiver *only* if the energy is tapped, and this is sensed by the transmitter by means of responding to electromagnetic near fields emanating from the receiver. Again, this is the same principle that applies in induction coupled devices, such as a transformer, which draws more power at the primary circuit, if power is drawn from the secondary circuit. This is different with the far field, which constantly draws the same energy from the transmitter, whether it is immediately received, or not.

The amplitude of other components (non-radiative/non-dipole) of the electromagnetic field close to the antenna may be quite powerful, but, because of more rapid fall-off with distance than $1/r$ behavior, they do not radiate energy to infinite distances. Instead, their energies remain trapped in the region near the antenna, not drawing power from the transmitter unless they excite a receiver in the area close to the antenna. Thus, the near fields only transfer energy to very nearby receivers, and, when they do, the result is felt as an extra power draw in the transmitter. As an example of such an effect, power is transferred across space in a common transformer or metal detector by means of near-field phenomena (in this case inductive coupling), in a strictly "short-range" effect (i.e., the range within one wavelength of the signal).

Classical EM Modelling

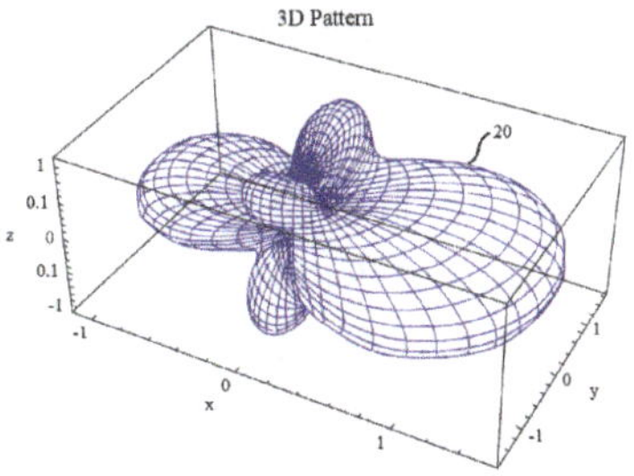

A "radiation pattern" for an antenna, by definition showing only the far field.

Solving Maxwell's equations for the electric and magnetic fields for a localized oscillating source, such as an antenna, surrounded by a homogeneous material (typically vacuum or air), yields fields that, far away, decay in proportion to $1/r$ where r is the distance from the source. These are the *radiating* fields, and the region where r is large enough for these fields to dominate is the *far field.*

In general, the fields of a source in a homogeneous isotropic medium can be written as a multipole expansion. The terms in this expansion are spherical harmonics (which give the angular dependence) multiplied by spherical Bessel functions (which give the radial dependence). For large r, the spherical Bessel functions decay as $1/r$, giving the radiated field above. As one gets closer and closer to the source (smaller r), approaching the *near field*, other powers of r become significant.

The next term that becomes significant is proportional to $1/r^2$ and is sometimes called the *induction term*. It can be thought of as the primarily magnetic energy stored in the field, and returned to the antenna in every half-cycle, through self-induction. For even smaller r, terms proportional to $1/r^3$ become significant; this is sometimes called the *electrostatic field term* and can be thought of as stemming from the electrical charge in the antenna element.

Very close to the source, the multipole expansion is less useful (too many terms are required for an accurate description of the fields). Rather, in the near field, it is sometimes useful to express the contributions as a sum of radiating fields combined with evanescent fields, where the latter are exponentially decaying with r. And in the source itself, or as soon as one enters a region of inhomogeneous materials, the multipole expansion is no longer valid and the full solution of Maxwell's equations is generally required.

Antennas

If an oscillating electrical current is applied to a conductive structure of some type, electric and magnetic fields will appear in space about that structure. If those fields extend some distance into space the structure is often termed an antenna. Such an antenna can be an assemblage of conductors in space typical of radio devices or it can be an aperture with a given current distribution radiating into space as is typical of microwave or optical devices. The actual values of the fields in space about the antenna are usually quite complex and can vary with distance from the antenna in various ways.

However, in many practical applications, one is interested only in effects where the distance from the antenna to the observer is very much greater than the largest dimension of the transmitting antenna. The equations describing the fields created about the antenna can be simplified by assuming a large separation and dropping all terms that provide only minor contributions to the final field. These simplified distributions have been termed the "far field" and usually have the property that the angular distribution of energy does not change with distance, although the energy levels still vary with distance and time. Such an angular energy distribution is usually termed an antenna pattern.

Note that, by the principle of reciprocity, the pattern observed when a particular antenna is transmitting is identical to the pattern measured when the same antenna is used for reception. Typically one finds simple relations describing the antenna far-field patterns, often involving trigonometric functions or at worst Fourier or Hankel transform relationships between the antenna current distributions and the observed far-field patterns. While far-field simplifications are very useful in engineering calculations, this does not mean the near-field functions cannot be calculated, especially using modern computer techniques. An examination of how the near fields form about an antenna structure can give great insight into the operations of such devices.

Impedance

The electromagnetic field in the far-field region of an antenna is independent of the details of the near field and the nature of the antenna. The wave impedance is the ratio of the strength of the electric and magnetic fields, which in the far field are in phase with each other. Thus, the far field "impedance of free space" is resistive and is given by:

$$Z_0 \stackrel{\text{def}}{=} \mu_0 c_0 = \sqrt{\frac{\mu_0}{\varepsilon_0}} = \frac{1}{\varepsilon_0 c_0}$$

With the usual approximation for the speed of light in free space $c_0 = 3 \times 10^8$ m/s gives the frequently used expression:

$$Z_0 \approx 120\pi \approx 377\ \Omega$$

The electromagnetic field in the near-field region of an electrically small coil antenna is predominantly magnetic. For small values of r/λ, the wave impedance of an inductor is low and inductive, at short range being asymptotic to:

$$|Z_w| \approx 240\pi^2 \frac{r}{\lambda} \approx 2370 \frac{r}{\lambda}$$

The electromagnetic field in the near-field region of an electrically short rod antenna is predominantly electric. For small values of r/λ, the wave impedance is high and capacitive, at short range being asymptotic to:

$$|Z_W| \approx 60 \frac{\lambda}{r}$$

In both cases, the wave impedance converges on that of free space as the range approaches the far field.

Quantum Field Theory View

In the quantum view of electromagnetic interactions, far-field effects are manifestations of real photons, whereas near-field effects are due to a mixture of real and virtual photons. Virtual photons composing near-field fluctuations and signals have effects that are of far shorter range than those of real photons.

Wave–particle Duality

Wave–particle duality is the concept that every elementary particle or quantic entity may be partly described in terms not only of particles, but also of waves. It expresses the inability of the classical concepts "particle" or "wave" to fully describe the behavior of quantum-scale objects. As Albert Einstein wrote: "*It seems as though we must use sometimes the one theory and sometimes the other, while at times we may use either. We are faced with a new kind of difficulty. We have two contradictory pictures of reality; separately neither of them fully explains the phenomena of light, but together they do.*"

Through the work of Max Planck, Einstein, Louis de Broglie, Arthur Compton, Niels Bohr and many others, current scientific theory holds that all particles also have a wave nature (and vice versa). This phenomenon has been verified not only for elementary particles, but also for compound particles like atoms and even molecules. For macroscopic particles, because of their extremely short wavelengths, wave properties usually cannot be detected.

Although the use of the wave-particle duality has worked well in physics, the *meaning* or *interpretation* has not been satisfactorily resolved.

Niels Bohr regarded the "duality paradox" as a fundamental or metaphysical fact of nature. A given kind of quantum object will exhibit sometimes wave, sometimes particle, character, in respectively different physical settings. He saw such duality as one aspect of the concept of complementarity. Bohr regarded renunciation of the cause-effect relation, or complementarity, of the space-time picture, as essential to the quantum mechanical account.

Werner Heisenberg considered the question further. He saw the duality as present for all quantic entities, but not quite in the usual quantum mechanical account considered by Bohr. He saw it in what is called second quantization, which generates an entirely new concept of fields which exist in ordinary space-time, causality still being visualizable. Classical field values (e.g. the electric and magnetic field strengths of Maxwell) are replaced by an entirely new kind of field value, as considered in quantum field theory. Turning the reasoning around, ordinary quantum mechanics can be deduced as a specialized consequence of quantum field theory.

Brief History of Wave and Particle Viewpoints

Democritus—the original *atomist*—argued that all things in the universe, including light, are composed of indivisible sub-components (light being some form of solar atom). At the beginning of the 11th Century, the Arabic scientist Alhazen wrote the first comprehensive treatise on optics; describing refraction, reflection, and the operation of a pinhole lens via rays of light traveling from the point of emission to the eye. He asserted that these rays were composed of particles of light. In 1630, René Descartes popularized and accredited the opposing wave description in his treatise on light, showing that the behavior of light could be re-created by modeling wave-like disturbances in a universal medium ("plenum"). Beginning in 1670 and progressing over three decades, Isaac Newton developed and championed his corpuscular hypothesis, arguing that the perfectly straight lines of reflection demonstrated light's particle nature; only particles could travel in such straight lines. He explained refraction by positing that particles of light accelerated laterally upon entering a denser medium. Around the same time, Newton's contemporaries Robert Hooke and Christiaan Huygens—and later Augustin-Jean Fresnel—mathematically refined the wave viewpoint, showing that if light traveled at different speeds in different media (such as water and air), refraction could be easily explained as the medium-dependent propagation of light waves. The resulting Huygens–Fresnel principle was extremely successful at reproducing light's behavior and was subsequently supported by Thomas Young's 1803 discovery of double-slit interference. The wave view did not immediately displace the ray and particle view, but began to dominate scientific thinking about light in the mid 19th century, since it could explain polarization phenomena that the alternatives could not.

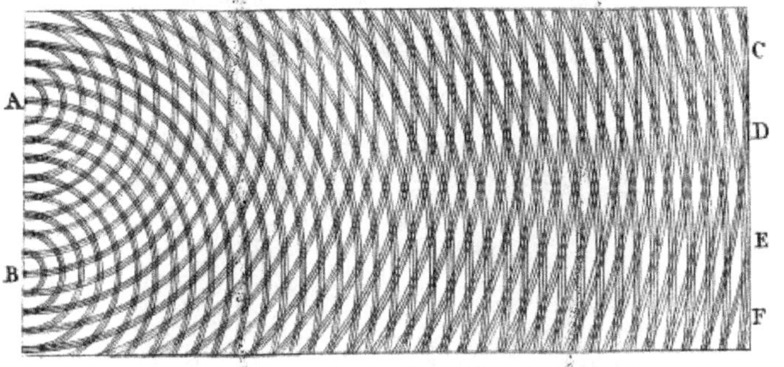

Thomas Young's sketch of two-slit diffraction of waves, 1803

James Clerk Maxwell discovered that he could apply his equations for electromagnetism, which had been previously discovered, along with a slight modification to describe self-propagating waves of oscillating electric and magnetic fields. When the propagation speed of these electromagnetic waves was calculated, the speed of light fell out. It quickly became apparent that visible light, ultraviolet light, and infrared light (phenomena thought previously to be unrelated) were all electromagnetic waves of differing frequency. The wave theory had prevailed—or at least it seemed to.

While the 19th century had seen the success of the wave theory at describing light, it had also witnessed the rise of the atomic theory at describing matter. Antoine Lavoisier deduced the law of conservation of mass and categorized many new chemical elements and compounds; and Joseph Louis Proust advanced chemistry towards the atom by showing that elements combined in definite proportions. This led John Dalton to propose that elements were invisible sub components; Amedeo Avogadro discovered diatomic gases and completed the basic atomic theory, allowing the correct molecular formulae of most known compounds—as well as the correct weights of atoms—to be deduced and categorized in a consistent manner. Dimitri Mendeleev saw an order in recurring chemical properties, and created a table presenting the elements in unprecedented order and symmetry.

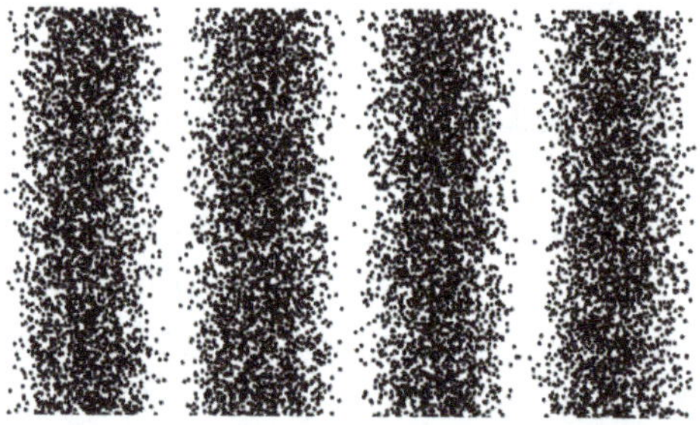

Particle impacts make visible the interference pattern of waves.

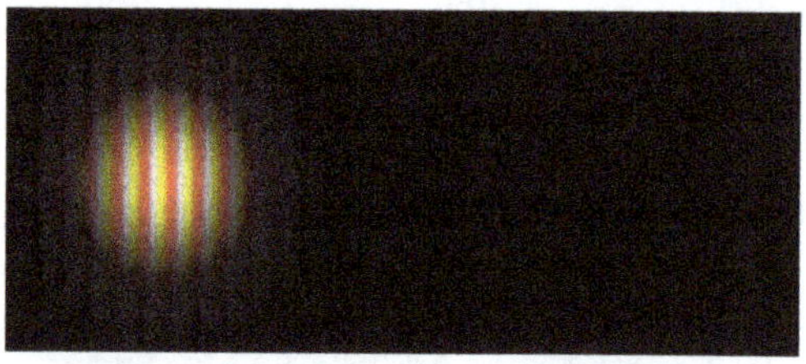

A quantum particle is represented by a wave packet.

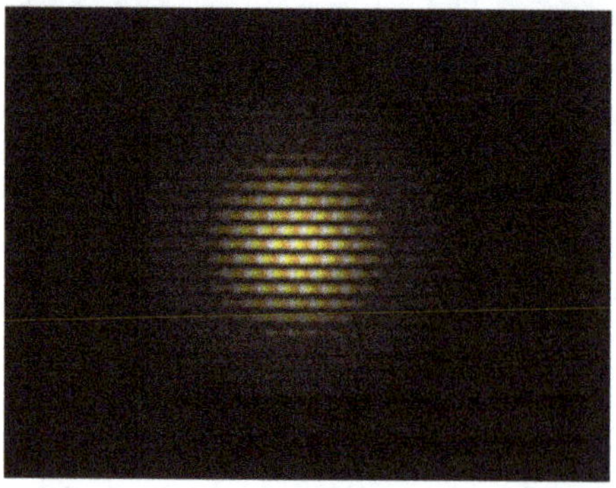

Interference of a quantum particle with itself.

Turn of the 20th Century and the Paradigm Shift

Particles of Electricity

At the close of the 19th century, the reductionism of atomic theory began to advance into the atom itself; determining, through physics, the nature of the atom and the operation of chemical reactions. Electricity, first thought to be a fluid, was now understood to consist of particles called electrons. This was first demonstrated by J. J. Thomson in 1897 when, using a cathode ray tube, he found that an electrical charge would travel across a vacuum (which would possess infinite resistance in classical theory). Since the vacuum offered no medium for an electric fluid to travel, this discovery could only be explained via a particle carrying a negative charge and moving through the vacuum. This *electron* flew in the face of classical electrodynamics, which had successfully treated electricity as a fluid for many years (leading to the invention of batteries, electric motors, dynamos, and arc lamps). More importantly, the intimate relation between electric charge and electromagnetism had been well documented following the discoveries of Michael Faraday and James Clerk Maxwell. Since electromagnetism was *known* to be a wave generated by a changing electric or magnetic *field* (a continuous, wave-like entity itself) an atomic/particle description of electricity and charge was a non sequitur. Furthermore, classical electrodynamics was not the only classical theory rendered incomplete.

Radiation Quantization

In 1901, Max Planck published an analysis that succeeded in reproducing the observed spectrum of light emitted by a glowing object. To accomplish this, Planck had to make an ad hoc mathematical assumption of quantized energy of the oscillators (atoms of the black body) that emit radiation. Einstein later proposed that electromagnetic radiation itself is quantized, not the energy of radiating atoms.

Black-body radiation, the emission of electromagnetic energy due to an object's heat, could not be explained from classical arguments alone. The equipartition theorem of classical mechanics, the basis of all classical thermodynamic theories, stated that an object's energy is partitioned equally among the object's vibrational modes. But applying the same reasoning to the electromagnetic emission of such a thermal object was not so successful. That thermal objects emit light had been long known. Since light was known to be waves of electromagnetism, physicists hoped to describe this emission via classical laws. This became known as the black body problem. Since the equipartition theorem worked so well in describing the vibrational modes of the thermal object itself, it was natural to assume that it would perform equally well in describing the radiative emission of such objects. But a problem quickly arose: if each mode received an equal partition of energy, the short wavelength modes would consume all the energy. This became clear when plotting the Rayleigh–Jeans law which, while correctly predicting the intensity of long wavelength emissions, predicted infinite total energy as the intensity diverges to infinity for short wavelengths. This became known as the ultraviolet catastrophe.

In 1900, Max Planck hypothesized that the frequency of light emitted by the black body depended on the frequency of the *oscillator* that emitted it, and the energy of these oscillators increased linearly with frequency (according to his constant *h*, where $E = h\nu$). This was not an unsound proposal considering that macroscopic oscillators operate similarly: when studying five simple harmonic oscillators of equal amplitude but different frequency, the oscillator with the highest frequency possesses the highest energy (though this relationship is not linear like Planck's). By demanding that high-frequency light must be emitted by an oscillator of equal frequency, and further requiring that this oscillator occupy higher energy than one of a lesser frequency, Planck avoided any catastrophe; giving an equal partition to high-frequency oscillators produced successively fewer oscillators and less emitted light. And as in the Maxwell–Boltzmann distribution, the low-frequency, low-energy oscillators were suppressed by the onslaught of thermal jiggling from higher energy oscillators, which necessarily increased their energy and frequency.

The most revolutionary aspect of Planck's treatment of the black body is that it inherently relies on an integer number of oscillators in thermal equilibrium with the electromagnetic field. These oscillators *give* their entire energy to the electromagnetic field, creating a quantum of light, as often as they are *excited* by the electromagnetic field, absorbing a quantum of light and beginning to oscillate at the corresponding frequency. Planck had intentionally created an atomic theory of the black body, but had unintentionally generated an atomic theory of light, where the black body never generates quanta of light at a given frequency with an energy less than hv. However, once realizing that he had quantized the electromagnetic field, he denounced particles of light as a limitation of his approximation, not a property of reality.

Photoelectric Effect Illuminated

While Planck had solved the ultraviolet catastrophe by using atoms and a quantized electromagnetic field, most contemporary physicists agreed that Planck's "light quanta" represented only flaws in his model. A more-complete derivation of black body radiation would yield a fully continuous and 'wave-like' electromagnetic field with no quantization. However, in 1905 Albert Einstein took Planck's black body model to produce his solution to another outstanding problem of the day: the photoelectric effect, wherein electrons are emitted from atoms when they absorb energy from light. Since their discovery eight years previously, electrons had been studied in physics laboratories worldwide.

In 1902 Philipp Lenard discovered that the energy of these ejected electrons did *not* depend on the intensity of the incoming light, but instead on its *frequency*. So if one shines a little low-frequency light upon a metal, a few low energy electrons are ejected. If one now shines a very intense beam of low-frequency light upon the same metal, a whole slew of electrons are ejected; however they possess the same low energy, there are merely *more of them*. The more light there is, the more electrons are ejected. Whereas in order to get high energy electrons, one must illuminate the metal with high-frequency light. Like blackbody radiation, this was at odds with a theory invoking continuous transfer of energy between radiation and matter. However, it can still be explained using a fully classical description of light, as long as matter is quantum mechanical in nature.

If one used Planck's energy quanta, and demanded that electromagnetic radiation at a given frequency could only transfer energy to matter in integer multiples of an energy quantum hν, then the photoelectric effect could be explained very simply. Low-frequency light only ejects low-energy electrons because each electron is excited by the absorption of a single photon. Increasing the intensity of the low-frequency light (increasing the number of photons) only increases the number of excited electrons, not their energy, because the energy of each photon remains low. Only by increasing the frequency of the light, and thus increasing the energy of the photons, can one eject electrons with higher energy. Thus, using Planck's constant *h* to determine the energy of the photons based upon their frequency, the energy of ejected electrons should also increase linearly with frequency; the gradient of the line being Planck's constant. These results were not confirmed until 1915, when Robert Andrews Millikan, who had previously determined the charge of the electron, produced experimental results in perfect accord with Einstein's predictions. While the energy of ejected electrons reflected Planck's constant, the existence of photons was not explicitly proven until the discovery of the photon antibunching effect, of which a modern experiment can be performed in undergraduate-level labs. This phenomenon could only be explained via photons, and not through any semi-classical theory (which could alternatively explain the photoelectric effect). When Einstein received his Nobel Prize in 1921, it was not for his more difficult and mathematically laborious special and general relativity, but for the simple, yet totally revolutionary, suggestion of quantized light. Einstein's "light quanta" would not be called photons until 1925, but even in 1905 they represented the quintessential example of wave-particle duality. Electromagnetic radiation propagates following linear wave equations, but can only be emitted or absorbed as discrete elements, thus acting as a wave and a particle simultaneously.

Einstein's Explanation of the Photoelectric Effect

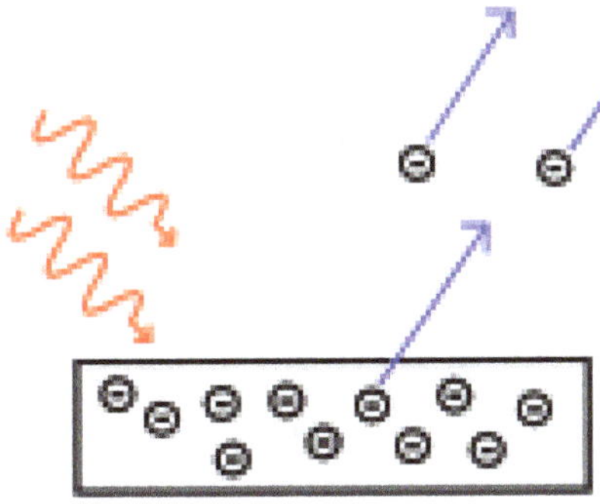

The photoelectric effect. Incoming photons on the left strike a metal plate (bottom), and eject electrons, depicted as flying off to the right.

In 1905, Albert Einstein provided an explanation of the photoelectric effect, a hitherto troubling experiment that the wave theory of light seemed incapable of explaining. He did so by postulating the existence of photons, quanta of light energy with particulate qualities.

In the photoelectric effect, it was observed that shining a light on certain metals would lead to an electric current in a circuit. Presumably, the light was knocking electrons out of the metal, causing current to flow. However, using the case of potassium as an example, it was also observed that while a dim blue light was enough to cause a current, even the strongest, brightest red light available with the technology of the time caused no current at all. According to the classical theory of light and matter, the strength or amplitude of a light wave was in proportion to its brightness: a bright light should have been easily strong enough to create a large current. Yet, oddly, this was not so.

Einstein explained this conundrum by postulating that the electrons can receive energy from electromagnetic field only in discrete portions (quanta that were called photons): an amount of energy E that was related to the frequency f of the light by

$$E = hf$$

where h is Planck's constant (6.626×10^{-34} J seconds). Only photons of a high enough frequency (above a certain *threshold* value) could knock an electron free. For example, photons of blue light had sufficient energy to free an electron from the metal, but photons of red light did not. One photon of light above the threshold frequency could release only one electron; the higher the frequency of a photon, the higher the kinetic energy of the emitted electron, but no amount of light (using technology available at the time) below the threshold frequency could release an electron. To "violate" this law would require extremely high-intensity lasers which had not yet been invented. Intensity-dependent phenomena have now been studied in detail with such lasers.

Einstein was awarded the Nobel Prize in Physics in 1921 for his discovery of the law of the photoelectric effect.

De Broglie's Wavelength

In 1924, Louis-Victor de Broglie formulated the de Broglie hypothesis, claiming that *all* matter, not just light, has a wave-like nature; he related wavelength (denoted as λ), and momentum (denoted as p):

$$\lambda = \frac{h}{p}$$

This is a generalization of Einstein's equation above, since the momentum of a photon is given by $p = \frac{E}{c}$ and the wavelength (in a vacuum) by $\lambda = \frac{c}{f}$, where c is the speed of light in vacuum.

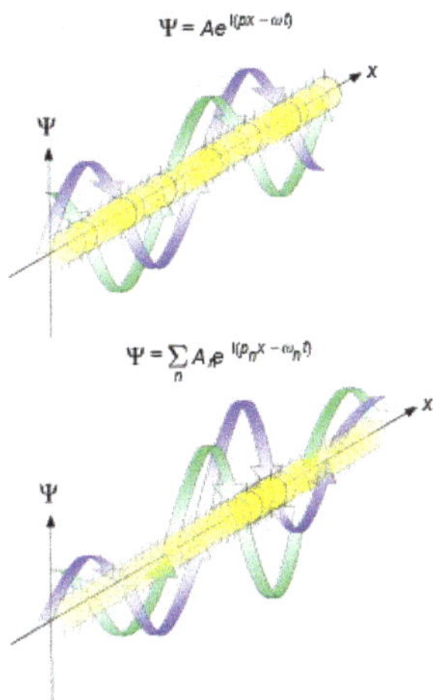

Propagation of de Broglie waves in 1d—real part of the complex amplitude is blue, imaginary part is green. The probability (shown as the colour opacity) of finding the particle at a given point x is spread out like a waveform; there is no definite position of the particle. As the amplitude increases above zero the curvature decreases, so the amplitude decreases again, and vice versa—the result is an alternating amplitude: a wave. Top: Plane wave. Bottom: Wave packet.

De Broglie's formula was confirmed three years later for electrons (which differ from photons in having a rest mass) with the observation of electron diffraction in two independent experiments. At the University of Aberdeen, George Paget Thomson passed a beam of electrons through a thin metal film and observed the predicted interference patterns. At Bell Labs, Clinton Joseph Davisson and Lester Halbert Germer guided their beam through a crystalline grid.

De Broglie was awarded the Nobel Prize for Physics in 1929 for his hypothesis. Thomson and Davisson shared the Nobel Prize for Physics in 1937 for their experimental work.

Heisenberg's Uncertainty Principle

In his work on formulating quantum mechanics, Werner Heisenberg postulated his uncertainty principle, which states:

$$\Delta x \Delta p \geq \frac{\hbar}{2}$$

where

Δ here indicates standard deviation, a measure of spread or uncertainty;

x and p are a particle's position and linear momentum respectively.

$\hbar$ is the reduced Planck's constant (Planck's constant divided by 2 π).

Heisenberg originally explained this as a consequence of the process of measuring:

Measuring position accurately would disturb momentum and vice versa, offering an example (the "gamma-ray microscope") that depended crucially on the de Broglie hypothesis. The thought is now, however, that this only partly explains the phenomenon, but that the uncertainty also exists in the particle itself, even before the measurement is made.

In fact, the modern explanation of the uncertainty principle, extending the Copenhagen interpretation first put forward by Bohr and Heisenberg, depends even more centrally on the wave nature of a particle: Just as it is nonsensical to discuss the precise location of a wave on a string, particles do not have perfectly precise positions; likewise, just as it is nonsensical to discuss the wavelength of a "pulse" wave traveling down a string, particles do not have perfectly precise momenta (which corresponds to the inverse of wavelength). Moreover, when position is relatively well defined, the wave is pulse-like and has a very ill-defined wavelength (and thus momentum). And conversely, when momentum (and thus wavelength) is relatively well defined, the wave looks long and sinusoidal, and therefore it has a very ill-defined position.

De Broglie–Bohm Theory

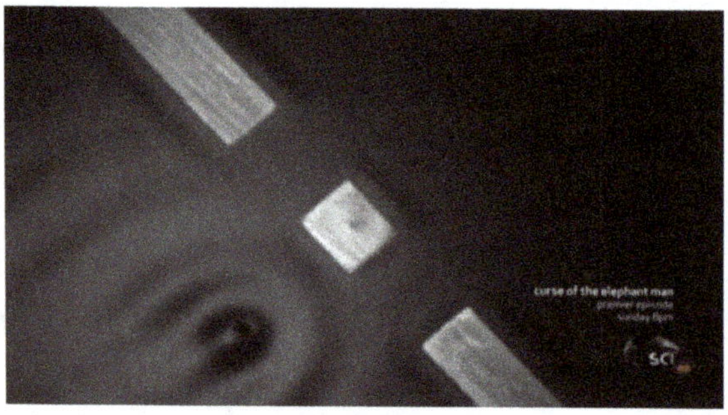

Couder experiments, "materializing" the *pilot wave* model.

De Broglie himself had proposed a pilot wave construct to explain the observed wave-particle duality. In this view, each particle has a well-defined position and momentum, but is guided by a wave function derived from Schrödinger's equation. The pilot wave theory was initially rejected because it generated non-local effects when applied to systems involving more than one particle. Non-locality, however, soon became established as an integral feature of quantum theory, and David Bohm extended de Broglie's model to explicitly include it.

In the resulting representation, also called the de Broglie–Bohm theory or Bohmian mechanics, the wave-particle duality vanishes, and explains the wave behaviour as a scattering with wave appearance, because the particle's motion is subject to a guiding equation or quantum potential. *"This idea seems to me so natural and simple, to resolve the wave-particle dilemma in such a clear and ordinary way, that it is a great mystery to me that it was so generally ignored"*, J.S.Bell.

The best illustration of the *pilot-wave model* was given by Couder's 2010 "walking droplets" experiments, demonstrating the pilot-wave behaviour in a macroscopic mechanical analog.

Wave Behavior of Large Objects

Since the demonstrations of wave-like properties in photons and electrons, similar experiments have been conducted with neutrons and protons. Among the most famous experiments are those of Estermann and Otto Stern in 1929. Authors of similar recent experiments with atoms and molecules, described below, claim that these larger particles also act like waves. A wave is basically a group of particles which moves in a particular form of motion, i.e. to and fro. If we break that flow by an object it will convert into radiants.

A dramatic series of experiments emphasizing the action of gravity in relation to wave–particle duality was conducted in the 1970s using the neutron interferometer. Neutrons, one of the components of the atomic nucleus, provide much of the mass of a nucleus and thus of ordinary matter. In the neutron interferometer, they act as quantum-mechanical waves directly subject to the force of gravity. While the results were not surprising since gravity was known to act on everything, including light, the self-interference of the quantum mechanical wave of a massive fermion in a gravitational field had never been experimentally confirmed before.

In 1999, the diffraction of C_{60} fullerenes by researchers from the University of Vienna was reported. Fullerenes are comparatively large and massive objects, having an atomic mass of about 720 u. The de Broglie wavelength of the incident beam was about 2.5 pm, whereas the diameter of the molecule is about 1 nm, about 400 times larger. In 2012, these far-field diffraction experiments could be extended to phthalocyanine molecules and their heavier derivatives, which are composed of 58 and 114 atoms respectively. In these experiments the build-up of such interference patterns could be recorded in real time and with single molecule sensitivity.

In 2003, the Vienna group also demonstrated the wave nature of tetraphenylporphyrin—a flat biodye with an extension of about 2 nm and a mass of 614 u. For this demonstration they employed a near-field Talbot Lau interferometer. In the same interferometer they also found interference fringes for $C_{60}F_{48.}$, a fluorinated buckyball with a mass of about 1600 u, composed of 108 atoms. Large molecules are already so complex that they give experimental access to some aspects of the quantum-classical interface, i.e., to certain decoherence mechanisms. In 2011, the interference of molecules as heavy as 6910 u could be demonstrated in a Kapitza–Dirac–Talbot–Lau interferometer. In 2013, the interference of molecules beyond 10,000 u has been demonstrated.

Whether objects heavier than the Planck mass (about the weight of a large bacterium) have a de Broglie wavelength is theoretically unclear and experimentally unreachable; above the Planck mass a particle's Compton wavelength would be smaller than the Planck length and its own Schwarzschild radius, a scale at which current theories of physics may break down or need to be replaced by more general ones.

Recently Couder, Fort, *et al.* showed that we can use macroscopic oil droplets on a vibrating surface as a model of wave–particle duality—localized droplet creates periodical waves around and interaction with them leads to quantum-like phenomena: interference in double-slit experiment, unpredictable tunneling (depending in complicated way on practically hidden state of field), orbit quantization (that particle has to 'find a resonance' with field perturbations it creates—after one orbit, its internal phase has to return to the initial state) and Zeeman effect.

Treatment in Modern Quantum Mechanics

Wave–particle duality is deeply embedded into the foundations of quantum mechanics. In the formalism of the theory, all the information about a particle is encoded in its *wave function*, a complex-valued function roughly analogous to the amplitude of a wave at each point in space. This function evolves according to a differential equation (generically called the Schrödinger equation). For particles with mass this equation has solutions that follow the form of the wave equation. Propagation of such waves leads to wave-like phenomena such as interference and diffraction. Particles without mass, like photons, have no solutions of the Schrödinger equation so have another wave.

The particle-like behavior is most evident due to phenomena associated with measurement in quantum mechanics. Upon measuring the location of the particle, the particle will be forced into a more localized state as given by the uncertainty principle. When viewed through this formalism, the measurement of the wave function will randomly "collapse", or rather "decohere", to a sharply peaked function at some location. For particles with mass the likelihood of detecting the particle at any particular location is equal to the squared amplitude of the wave function there. The measurement will return a well-defined position, (subject to uncertainty), a property traditionally associated with particles. It is important to note that a measurement is only a particular type of interaction where some data is recorded and the measured quantity is forced into a particular eigenstate. The act of measurement is therefore not fundamentally different from any other interaction.

Following the development of quantum field theory the ambiguity disappeared. The field permits solutions that follow the wave equation, which are referred to as the wave functions. The term particle is used to label the irreducible representations of the Lorentz group that are permitted by the field. An interaction as in a Feynman diagram is accepted as a calculationally convenient approximation where the outgoing legs are known to be simplifications of the propagation and the internal lines are for some order in an expansion of the field interaction. Since the field is non-local and quantized, the phenomena which previously were thought of as paradoxes are explained. Within the limits of the wave-particle duality the quantum field theory gives the same results.

Visualization

There are two ways to visualize the wave-particle behaviour: by the "standard model", described below; and by the Broglie–Bohm model, where no duality is perceived.

Below is an illustration of wave–particle duality as it relates to De Broglie's hypothesis and Heisenberg's uncertainty principle (above), in terms of the position and momentum space wavefunctions for one spinless particle with mass in one dimension. These wavefunctions are Fourier transforms of each other.

The more localized the position-space wavefunction, the more likely the particle is to be found with the position coordinates in that region, and correspondingly the momentum-space wavefunction is less localized so the possible momentum components the particle could have are more widespread.

Conversely the more localized the momentum-space wavefunction, the more likely the particle is to be found with those values of momentum components in that region, and correspondingly the less localized the position-space wavefunction, so the position coordinates the particle could occupy are more widespread.

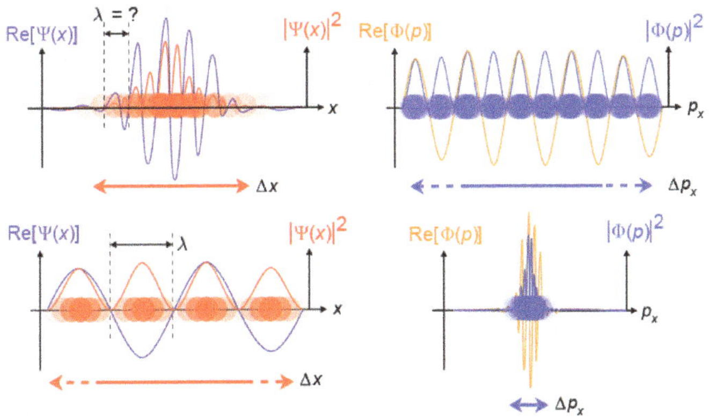

Position x and momentum p wavefunctions corresponding to quantum particles. The colour opacity (%) of the particles corresponds to the probability density of finding the particle with position x or momentum component p. Top: If wavelength λ is unknown, so are momentum p, wave-vector k and energy E (de Broglie relations). As the particle is more localized in position space, Δx is smaller than for Δp_x. Bottom: If λ is known, so are p, k, and E. As the particle is more localized in momentum space, Δp is smaller than for Δx.

Alternative Views

Wave–particle duality is an ongoing conundrum in modern physics. Most physicists accept wave-particle duality as the best explanation for a broad range of observed phenomena; however, it is not without controversy. Alternative views are also presented here. These views are not generally accepted by mainstream physics, but serve as a basis for valuable discussion within the community.

Both-particle-and-wave View

The pilot wave model, originally developed by Louis de Broglie and further developed by David Bohm into the hidden variable theory proposes that there is no duality, but rather a system exhibits both particle properties and wave properties simultaneously, and particles are guided, in a deterministic fashion, by the pilot wave (or its "quantum potential") which will direct them to areas of constructive interference in preference to areas of destructive interference. This idea is held by a significant minority within the physics community.

At least one physicist considers the "wave-duality" as not being an incomprehensible mystery. L.E. Ballentine, *Quantum Mechanics, A Modern Development*, p. 4, explains:

When first discovered, particle diffraction was a source of great puzzlement. Are "particles" really "waves?" In the early experiments, the diffraction patterns were detected holistically by means of a photographic plate, which could not detect individual particles. As a result, the notion grew that particle and wave properties were mutually incompatible, or complementary, in the sense that different measurement apparatuses would be required to observe them. That idea, however, was only an unfortunate generalization from a technological limitation. Today it is possible to detect the arrival of individual electrons, and to see the diffraction pattern emerge as a statistical pattern made up of many small spots (Tonomura et al., 1989). Evidently, quantum particles are indeed particles, but whose behaviour is very different from classical physics would have us to expect.

The Afshar experiment (2007) may suggest that it is possible to simultaneously observe both wave and particle properties of photons. This claim is, however, disputed by other scientists.

Wave-only View

At least one scientist proposes that the duality can be replaced by a "wave-only" view. In his book *Collective Electrodynamics: Quantum Foundations of Electromagnetism* (2000), Carver Mead purports to analyze the behavior of electrons and photons purely in terms of electron wave functions, and attributes the apparent particle-like behavior to quantization effects and eigenstates. According to reviewer David Haddon:

Mead has cut the Gordian knot of quantum complementarity. He claims that atoms, with their neutrons, protons, and electrons, are not particles at all but pure waves of matter. Mead cites as the gross evidence of the exclusively wave nature of both light and matter the discovery between 1933 and 1996 of ten examples of pure wave phenomena, including the ubiquitous laser of CD players, the self-propagating electrical currents of superconductors, and the Bose–Einstein condensate of atoms.

Albert Einstein, who, in his search for a Unified Field Theory, did not accept wave-particle duality, wrote:

This double nature of radiation (and of material corpuscles)...has been interpreted by quantum-mechanics in an ingenious and amazingly successful fashion. This interpretation...appears to me as only a temporary way out...

The many-worlds interpretation (MWI) is sometimes presented as a waves-only theory, including by its originator, Hugh Everett who referred to MWI as "the wave interpretation".

The *Three Wave Hypothesis* of R. Horodecki relates the particle to wave. The hypothesis implies that a massive particle is an intrinsically spatially as well as temporally extended wave phenomenon by a nonlinear law.

Particle-only View

Still in the days of the old quantum theory, a pre-quantum-mechanical version of wave–particle duality was pioneered by William Duane, and developed by others including Alfred Landé. Duane explained diffraction of x-rays by a crystal in terms solely of their particle aspect. The deflection of the trajectory of each diffracted photon was explained as due to quantized momentum transfer from the spatially regular structure of the diffracting crystal.

Neither-wave-nor-particle View

It has been argued that there are never exact particles or waves, but only some compromise or intermediate between them. For this reason, in 1928 Arthur Eddington coined the name "*wavicle*" to describe the objects although it is not regularly used today. One consideration is that zero-dimensional mathematical points cannot be observed. Another is that the formal representation of such points, the Dirac delta function is unphysical, because it cannot be normalized. Parallel arguments apply to pure wave states. Roger Penrose states:

"Such 'position states' are idealized wavefunctions in the opposite sense from the momentum states. Whereas the momentum states are infinitely spread out, the position states are infinitely concentrated. Neither is normalizable [...]."

Relational Approach to wave–particle Duality

Relational quantum mechanics is developed which regards the detection event as establishing a relationship between the quantized field and the detector. The inherent ambiguity associated with applying Heisenberg's uncertainty principle and thus wave–particle duality is subsequently avoided.

Applications

Although it is difficult to draw a line separating wave–particle duality from the rest of quantum mechanics, it is nevertheless possible to list some applications of this basic idea.

- Wave–particle duality is exploited in electron microscopy, where the small wavelengths associated with the electron can be used to view objects much smaller than what is visible using visible light.
- Similarly, neutron diffraction uses neutrons with a wavelength of about 0.1 nm, the typical spacing of atoms in a solid, to determine the structure of solids.
- Photos are now able to show this dual nature, perhaps this will lead to new ways of examining and recording this behaviour.

Uniqueness Theorems

Uniqueness Theorem 1

Given $\nabla . \vec{V} = s(\vec{x})$

and $\nabla \times \vec{V} = \vec{c}(\vec{x})$ (with $\nabla . \vec{c}(\vec{x}) = 0$, of course)

if $\vec{\mathbf{V}}$ goes to zero at infinity (fast enough),

then $\vec{\mathbf{V}}$ can be uniquely written in terms of $s(\vec{x})$ and $\vec{c}(\vec{x})$.

Indeed the solution can be given:

$$V = -\nabla \quad (\vec{x}) + \nabla \times \vec{A}(\vec{x})$$

with

$$\Phi(\vec{x}) = \frac{1}{4\pi} \int \frac{s(\vec{x})}{|\vec{x} - \vec{x}'|} d^3\vec{x}'$$

$$\vec{A}(\vec{x}) = \frac{1}{4\pi} \int \frac{\vec{c}(\vec{x})}{|\vec{x} - \vec{x}'|} d^3\vec{x}'$$

Proof of Uniqueness Theorem 1

Steps Involved:

- Show $\nabla.\vec{V} = s(\vec{x})$, using $\nabla^2(\frac{1}{r}) = -4\pi\delta(r)$
- Show $\nabla \times \vec{V} = \vec{c}(\vec{x})$ using integration by parts. You'll have to use the $\nabla.\vec{c}(\vec{x}) = 0$ everywhere, and $\vec{c}(\vec{x}) = 0$ at large distances (or goes to zero fast enough)

Unique Scalar, given $\Delta^2\phi$ and Boundary Conditions

Uniqueness Theorem 2

For a scalar $\phi(\vec{x})$

given $\nabla^2\phi$ everywhere,

and given on a closed surface $\phi(\vec{x})$ or $\nabla\phi.\hat{n}$ on a closed surface

a unique solution for $\phi(\vec{x})$ exists

Steps for Proving Uniqueness Theorem 2

- Consider two solutions ϕ_1 and ϕ_2, and define $\psi = \phi_1 - \phi_2$
- Using $\oint(\psi\nabla\psi).d\vec{S} = \int(\nabla\psi).(del\psi)dV + \int\psi\nabla^2\psi dV$

  Show that $|\nabla\psi| = 0$ everywhere in the enclosed volume

  (Use $\psi = 0$ or $\nabla\psi.\hat{n} = 0$ at the boundary)

- Note: the boundary conditions may be of the form $\psi = 0$ on some part of the boundary and $\nabla\psi = 0$ on the remaining part.

Unique vector, given $\nabla \times (\nabla \times \vec{A})$

Uniqueness Theorem 3 [for a Vector $\vec{A}(\vec{x})$]

Given $\nabla \times (\nabla \times \vec{A})$ everywhere,

and given $\vec{A} \times \hat{n}$ or $(\nabla \times \vec{A}) \times \hat{n}$ on a closed surface a unique solution for $\vec{A}(\vec{x})$ exists.

Steps for Proving Uniqueness Theorem 3

- Consider two solutions $\vec{A}_1$ and $\vec{A}_2$, and define $\vec{a} = \vec{A}_1 - \vec{A}_2$
- Using

$$\oint\left[\vec{a}\times(\nabla\times\vec{a})\right].d\vec{S} = \int(\nabla\times\vec{a}).(\nabla\times\vec{a})dV - \int\vec{a}.\left[\nabla\times(\nabla\times\vec{a})dV\right]$$

Show that $\left|\nabla\times\vec{a}\right| = 0$ everywhere in the enclosed volume

(Use $\vec{a}\times\hat{n} = 0$ or $(\nabla\times\vec{a})\times\hat{n} = 0$ at the boundary)

- One may have $\vec{a}\times\hat{n} = 0$ on some part of the boundary and $(\nabla\times\vec{a})\times\hat{n} = 0$ n the remainder.
- $\vec{a}\times\hat{n}$: tangential component of $\vec{a}$ to the surface

A Caution About $\nabla^2\vec{A}$

- $\nabla^2\vec{A} \neq \left(\nabla^2 A_x\right)\hat{x} + \left(\nabla^2 A_y\right)\hat{y} + \left(\nabla^2 A_z\right)\hat{z}$
- In fact $\nabla^2\vec{\mathbf{A}}$ is defined through

$$\nabla^2\vec{\mathbf{A}} = -\nabla\times\left(\nabla\times\vec{A}\right) + \nabla(\nabla.\vec{A})$$

Uniqueness Theorems: Applications

- If a solution is found by hook by or crook, we can be sure that this is the only solution
- A search for simple solutions, with certain symmetry properties, if successful, can solve the problem completely.
- Tricks like the method of images work.

Separation of Variables

When to Use

This technique works when there is some symmetry in the boundary conditions of the problem, which suggests the use of certain coordinates.

If the boundary conditions are of the form

- $\Phi(x=a)=\phi_0$, for all $(y,z) \Rightarrow$ cartesian coordinates
- $\Phi(r=a)=\phi_0$, for all $(\theta,\phi) \Rightarrow$ spherical polar coordinates
- $\Phi(r=a)=\phi_0$, for all $(z,\phi) \Rightarrow$ cylindrical coordinates

Cartesian Coordinate

$$\Phi(x,y,z)=X(x)\,Y(y)\,Z(z)$$

- Form of the solution:

$$X(x)=\begin{cases} Ae^{ik_x x}+Be^{-ik_x x} \\ Ae^{\kappa_x x}+Be^{-\kappa_x x} \end{cases}$$

Similarly for $Y(y)$ and $Z(z)$

- The solutions along x; y; z direction can be individually oscillatory $e^{\pm ikx}$ or hyperbolic $e^{\pm\kappa X}$.
- All three solutions cannot be propagating simultaneously, neither can all be hyperbolic at the same time.

Spherical Polar Coordinates

$$\Phi(r,\theta,\phi)=R_\ell(r)\Theta_\ell^m(\theta)\Phi_m(\phi)$$

Form of the Solution

- $R_\ell = A\ell r^\ell + B_\ell r^{-\ell-1}$
  - $A_\ell = 0$ if the solution is to be finite at infinity,
  - $B_\ell = 0$ if it is to be finite at the origin
- $\Theta_\ell^m(\theta)=C_\ell P_\ell^m(\cos\theta)+D_\ell Q_\ell^m(\cos\theta)$
  - P_ℓ^m, Q_ℓ^m : associated Legendre polynomials
  - $P_\ell^m = |Y_\ell^m|$, magnitudes of spherical harmonics
  - $D_\ell = 0$ if the solution is finite along z axis, since blows up there

- $\Phi(\phi)=\begin{cases} Ee^{im\phi}+Fe^{-im\phi} & (m \neq 0) \\ E\phi + F & (m=0) \end{cases}$
  - Azimuthal symmetry $\Rightarrow m=0$
  - $\oplus$ single-valued solution $\Rightarrow E=0$
  - $\mathrm{P}_\ell^0(\cos\theta)=P_\ell(\cos\theta)$, Legendre polynomials

Cylindrical Coordinates

$$\Phi(r,\phi,Z)=R_n(r)\Phi_n(\phi)Z(z)$$

Form of the Solution

- $R_n(r)=\begin{cases} \mathrm{A}_n J_n(kr)+\mathrm{B}_n \mathrm{N}_n(kr) & (\mathrm{k}\neq 0) \\ \mathrm{A}_n \mathrm{r}^n+\mathrm{B}_n \mathrm{r}^{-n} & (\mathrm{k}=0,\ \mathrm{n}\neq 0) \\ \mathrm{A}\ln r+\mathrm{B} & (\mathrm{k}=\mathrm{n}=0) \end{cases}$
  - J_n : Bessel functions, K_n : associated Bessel functions
  - $B_n=0$ if Φ is to be finite at the origin
- $\Phi_n(\phi)=\begin{cases} C_n e^{in\phi}+D_n \mathrm{e}^{-in\phi} & (n\neq 0) \\ C\phi+D & (n=0) \end{cases}$
  - Azimuthal symmetry $\Rightarrow \mathrm{n}=0$
  - $\oplus$ single-valued solution $\Rightarrow \mathrm{C}=0$
- $Z(z)=\begin{cases} Ee^{\mathrm{kz}}+Fe^{-\mathrm{kz}} & (\mathrm{k}\neq 0) \\ Ez+F & (\mathrm{k}=0) \end{cases}$
  - Z(z) can be oscillatory, in which case $R_n(r)$ involves modified Bessel functions

References

- Camilleri, K. (2009). Heisenberg and the Interpretation of Quantum Mechanics: the Physicist as Philosopher, Cambridge University Press, Cambridge UK, ISBN 978-0-521-88484-6
- Fort, E.; Eddi, A.; Boudaoud, A.; Moukhtar, J.; Couder, Y. (2010). "Path-memory induced quantization of classical orbits". PNAS. 107 (41): 17515–17520. doi:10.1073/pnas.1007386107
- J Rosen. "Redundancy and superfluity for electromagnetic fields and potentials". American Journal of Physics. 48 (12): 1071. Bibcode:1980AmJPh..48.1071R. doi:10.1119/1.12289

- Jean-Michel Lourtioz (2005-05-23). Photonic Crystals: Towards Nanoscale Photonic Devices. Berlin: Springer. p. 84. ISBN 3-540-24431-X
- Harrison, David (2002). "Complementarity and the Copenhagen Interpretation of Quantum Mechanics". UPSCALE. Dept. of Physics, U. of Toronto. Retrieved 2008-06-21
- Estermann, I.; Stern O. (1930). "Beugung von Molekularstrahlen". Zeitschrift für Physik. 61 (1-2): 95–125. Bibcode:1930ZPhy...61...95E. doi:10.1007/BF01340293
- B Jiang & J Wu & L.A. Povinelli (1996). "The Origin of Spurious Solutions in Computational Electromagnetics". Journal of Computational Physics. 125 (1): 104. Bibcode:1996JCoPh.125..104J. doi:10.1006/jcph.1996.0082
- S. F. Mahmoud (1991). Electromagnetic Waveguides: Theory and Applications. London UK: Institution of Electrical Engineers. Chapter 2. ISBN 0-86341-232-7
- Juffmann, Thomas; et al. (25 March 2012). "Real-time single-molecule imaging of quantum interference". Nature Nanotechnology. Retrieved 27 March 2012
- "Observing the quantum behavior of light in an undergraduate laboratory". American Journal of Physics. 72: 1210. Bibcode:2004AmJPh..72.1210T. doi:10.1119/1.1737397
- Horodecki, R. (1981). "De broglie wave and its dual wave". Phys. Lett. A. 87 (3): 95–97. Bibcode:-1981PhLA...87...95H. doi:10.1016/0375-9601(81)90571-5
- Taflove A & Hagness S C (2005). Computational Electrodynamics: The Finite-difference Time-domain Method. Boston MA: Artech House. Chapters 6 & 7. ISBN 1-58053-832-0
- Littlejohn, Robert (Fall 2007). "Gaussian, SI and Other Systems of Units in Electromagnetic Theory" (PDF). Physics 221A, University of California, Berkeley lecture notes. Retrieved 2008-05-06
- Penrose, Roger (2007). The Road to Reality: A Complete Guide to the Laws of the Universe. Vintage. p. 521, §21.10. ISBN 978-0-679-77631-4

2

Basics of Electrodynamics

The physical field that is produced by electrically charged objects is known as electromagnetic field. This section has been carefully written to provide an easy understanding of the varied facets of electrodynamics. This chapter explores the notions such as stationary states and plane waves and other concepts that form the basics of electrodynamics. The chapter strategically encompasses and incorporates the major components and key concepts of electromagnetic fields and waves, providing a complete understanding.

Relaxation (Physics)

In the physical sciences, relaxation usually means the return of a perturbed system into equilibrium. Each relaxation process can be categorized by a relaxation time τ. The simplest theoretical description of relaxation as function of time *t* is an exponential law exp(-*t*/τ).

Relaxation in Simple Linear Systems

Mechanics: Damped Unforced Oscillator

Let the homogeneous differential equation:

$$m\frac{d^2y}{dt^2}+\gamma\frac{dy}{dt}+ky=0$$

model damped unforced oscillations of a weight on a spring.

The displacement will then be of the form $y(t)=Ae^{-t/T}\cos(\mu t-\delta)$. The constant T is called the relaxation time of the system and the constant μ is the quasi-frequency.

Electronics: The RC Circuit

In an RC circuit containing a charged capacitor and a resistor, the voltage decays exponentially:

$$V(t)=V_0e^{-\frac{t}{RC}},$$

The constant $\tau = RC$ is called the *relaxation time* of the circuit. A nonlinear oscillator circuit which generates a repeating waveform by the repetitive discharge of a capacitor through a resistance is called a *relaxation oscillator*.

Relaxation in Condensed Matter Physics

In condensed matter physics, relaxation is usually studied as a linear response to a small external perturbation. Since the underlying microscopic processes are active even in the absence of external perturbations, one can also study "relaxation *in* equilibrium" instead of the usual "relaxation *into* equilibrium".

Stress Relaxation

In continuum mechanics, *stress relaxation* is the gradual disappearance of stresses from a viscoelastic medium after it has been deformed.

Dielectric Relaxation Time

In dielectric materials, the dielectric polarization P depends on the electric field E. If E changes, $P(t)$ reacts: the polarization *relaxes* towards a new equilibrium.

The dielectric relaxation time is closely related to the electrical conductivity. In a semiconductor it is a measure of how long it takes to become neutralized by conduction process. This relaxation time is small in metals and can be large in semiconductors and insulators.

Liquids and Amorphous Solids

An amorphous solid, such as amorphous indomethacin displays a temperature dependence of molecular motion, which can be quantified as the average relaxation time for the solid in a metastable supercooled liquid or glass to approach the molecular motion characteristic of a crystal. Differential scanning calorimetry can be used to quantify enthalpy change due to molecular structural relaxation.

The term "structural relaxation" was introduced in the scientific literature in 1947/48 without any explanation, applied to NMR, and meaning the same as "thermal relaxation".

Spin Relaxation in NMR

In nuclear magnetic resonance, relaxation is of prime importance.

Chemical Relaxation Methods

In chemical kinetics, relaxation methods are used for the measurement of very fast reaction rates. A system initially at equilibrium is perturbed by a rapid change in a parameter such as the temperature (most commonly), the pressure, the electric field or the pH of the solvent. The return to equilibrium is then observed, usually by spectroscopic means, and the relaxation time measured. In combination with the chemical equilibrium constant of the system, this enables the determination of the rate constants for the forward and reverse reactions.

Relaxation in Atmospheric Sciences

Desaturation of Clouds

Consider a supersaturated portion of a cloud. Then shut off the updrafts, entrainment, or any other vapor sources/sinks and things that would induce the growth of the particles (ice or water). Then wait for this supersaturation to reduce and become just saturation (relative humidity = 100%), which is the equilibrium state. The time it took for this to happen is called relaxation time. It will happen as ice crystals or liquid water content grow within the cloud and will thus consume the contained moisture. The dynamics of relaxation are very important in cloud physics modeling because if models do not take relaxation time into account, then it is highly probable that error will creep into the system.

In water clouds where the concentrations are larger (hundreds per cm^3) and the temperatures are warmer (thus allowing for much lower supersaturation rates as compared to ice clouds), the relaxation times will be very low (seconds to minutes).

In ice clouds the concentrations are lower (just a few per liter) and the temperatures are colder (very high supersaturation rates) and so the relaxation times can be hours and hours.

$\tau=(4\pi DNRK)^{-1}$

where

- D = diffusion coefficient [m2/s]
- N = concentration (of ice crystals or water droplets) [m−3]
- R = mean radius of particles [m]
- K = capacitance [unitless]

Relaxation in Astronomy

In astronomy, relaxation time relates to clusters of gravitationally interacting bodies, for instance, stars in a galaxy. The relaxation time is a measure of the time it takes for one object in the system (the "test star") to be significantly perturbed by other objects in the system (the "field stars"). It is most commonly defined as the time for the test star's velocity to change by of order itself.

Suppose that the test star has velocity v. As the star moves along its orbit, its motion will be randomly perturbed by the gravitational field of nearby stars. The relaxation time can be shown to be

$$T_r = \frac{0.34\sigma^3}{G^2 m\rho \ln \Lambda}$$

$$\approx 0.95\times10^{10}\left(\frac{\sigma}{200\text{kms}^{-1}}\right)^3\left(\frac{\rho}{10^6 M_\odot \text{pc}^{-3}}\right)^{-1}\left(\frac{m_\star}{M_\odot}\right)^{-1}\left(\frac{\ln \Lambda}{15}\right)^{-1} \text{yr}$$

where ρ is the mean density, m is the test-star mass, σ is the 1d velocity dispersion of the field stars, and $\ln \Lambda$ is the Coulomb logarithm.

Various events occur on timescales relating to the relaxation time, including core collapse, energy equipartition, and formation of a Bahcall-Wolf cusp around a supermassive black hole.

Stationary State

A stationary state is a pure quantum state with all observables independent of time. It is an eigenvector of the Hamiltonian. This corresponds to a state with a single definite energy (instead of a quantum superposition of different energies). It is also called energy eigenvector, energy eigenstate, energy eigenfunction, or energy eigenket. It is very similar to the concept of atomic orbital and molecular orbital in chemistry, with some slight differences explained below.

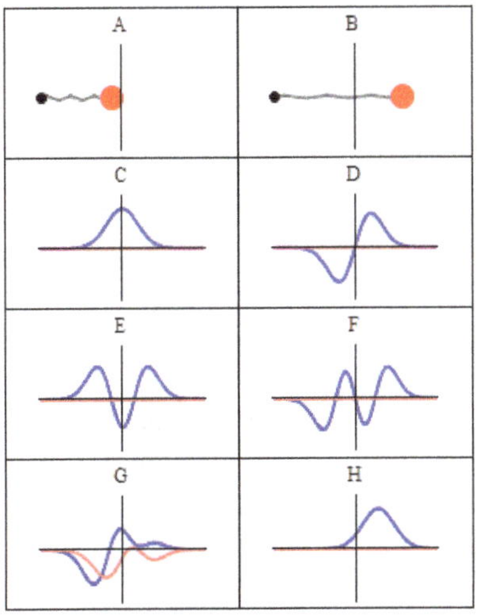

A harmonic oscillator in classical mechanics (A–B) and quantum mechanics (C–H). In (A–B), a ball, attached to a spring, oscillates back and forth. (C–H) are six solutions to the Schrödinger Equation for this situation. The horizontal axis is position, the vertical axis is the real part (blue) or imaginary part (red) of the wavefunction. (C,D,E,F), but not (G,H), are stationary states, or standing waves. The standing-wave oscillation frequency, times Planck's constant, is the energy of the state.

A stationary state is called *stationary* because the system remains in the same state as time elapses, in every observable way. For a single-particle Hamiltonian, this means that the particle has a constant probability distribution for its position, its velocity, its spin, etc. (This is true assuming the particle's environment is also static, i.e. the Hamiltonian is unchanging in time.) The wavefunction itself is not stationary: It continually changes its overall complex phase factor, so as to form a standing wave. The oscillation frequency of the standing wave, times Planck's constant, is the energy of the state according to the Planck–Einstein relation.

Stationary states are quantum states that are solutions to the time-independent Schrödinger equation:

$$\hat{H}|\Psi\rangle = E_{\Psi}|\Psi\rangle,$$

where

- $|\Psi\rangle$ is a quantum state, which is a stationary state if it satisfies this equation;
- $\hat{H}$ is the Hamiltonian operator;
- E_{Ψ} is a real number, and corresponds to the energy eigenvalue of the state $|\Psi\rangle$.

This is an eigenvalue equation: $\hat{H}$ is a linear operator on a vector space, $|\Psi\rangle$ is an eigenvector of $\hat{H}$, and E_{Ψ} is its eigenvalue.

If a stationary state $|\Psi\rangle$ is plugged into the time-dependent Schrödinger Equation, the result is:

$$i\hbar\frac{\partial}{\partial t}|\Psi\rangle = E_{\Psi}|\Psi\rangle$$

Assuming that $\hat{H}$ is time-independent (unchanging in time), this equation holds for any time *t*. Therefore, this is a differential equation describing how $|\Psi\rangle$ varies in time. Its solution is:

$$|\Psi(t)\rangle = e^{-iE_{\Psi}t/\hbar}|\Psi(0)\rangle$$

Therefore, a stationary state is a standing wave that oscillates with an overall complex phase factor, and its oscillation angular frequency is equal to its energy divided by $\hbar$.

Stationary State Properties

As shown, a stationary state is not mathematically constant:

$$|\Psi(t)\rangle = e^{-iE_{\Psi}t/\hbar}|\Psi(0)\rangle$$

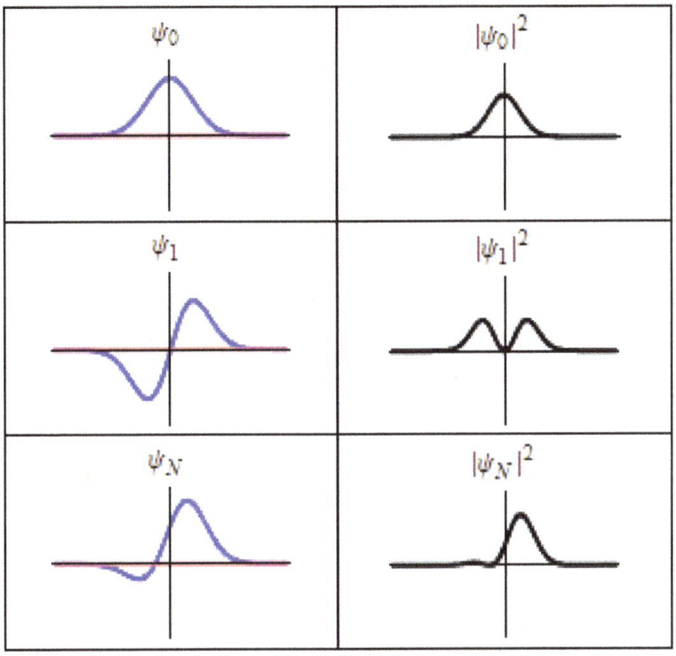

Three wavefunction solutions to the time-dependent Schrödinger equation for a harmonic oscillator. Left: The real part (blue) and imaginary part (red) of the wavefunction. Right: The probability of finding the particle at a certain position. The top two rows are two stationtary states, and the bottom is the superposition state $\psi_N \equiv (\psi_0 + \psi_1)/\sqrt{2}$, which is not a stationary state. The right column illustrates why stationary states are called "stationary".

However, all observable properties of the state are in fact constant. For example, if $|\Psi(t)\rangle$ represents a simple one-dimensional single-particle wavefunction $\Psi(x,t)$, the probability that the particle is at location x is:

$$|\Psi(x,t)|^2 = \left|e^{-iE_\psi t/\hbar}\Psi(x,0)\right|^2 = \left|e^{-iE_\psi t/\hbar}\right|^2 \left|\Psi(x,0)\right|^2 = \left|\Psi(x,0)\right|^2$$

which is independent of the time t.

The Heisenberg picture is an alternative mathematical formulation of quantum mechanics where stationary states are truly mathematically constant in time.

As mentioned above, these equations assume that the Hamiltonian is time-independent. This means simply that stationary states are only stationary when the rest of the system is fixed and stationary as well. For example, a 1s electron in a hydrogen atom is in a stationary state, but if the hydrogen atom reacts with another atom, then the electron will of course be disturbed.

Spontaneous Decay

Spontaneous decay complicates the question of stationary states. For example, according to simple (nonrelativistic) quantum mechanics, the hydrogen atom has many stationary states: 1s, 2s, 2p, and so on, are all stationary states. But in reality, only the ground state 1s is truly "stationary": An electron in a higher energy level will spontaneously emit one or more photons to decay into the ground state. This seems to contradict the idea that stationary states should have unchanging properties.

The explanation is that the Hamiltonian used in nonrelativistic quantum mechanics is only an approximation to the Hamiltonian from quantum field theory. The higher-energy electron states (2s, 2p, 3s, etc.) are stationary states according to the approximate Hamiltonian, but *not* stationary according to the true Hamiltonian, because of vacuum fluctuations. On the other hand, the 1s state is truly a stationary state, according to both the approximate and the true Hamiltonian.

Comparison to "Orbital" in Chemistry

An orbital is a stationary state (or approximation thereof) of a one-electron atom or molecule; more specifically, an atomic orbital for an electron in an atom, or a molecular orbital for an electron in a molecule.

For a molecule that contains only a single electron (e.g. atomic hydrogen or H_2^+), an orbital is exactly the same as a total stationary state of the molecule. However, for a many-electron molecule, an orbital is completely different from a total stationary state, which is a many-particle state requiring a more complicated description (such as a Slater determinant). In particular, in a many-electron molecule, an orbital is not the total stationary state of the molecule, but rather the stationary state of a single electron within the molecule. This concept of an orbital is only meaningful under the approximation that if we ignore the electron-electron repulsion terms in the Hamiltonian as a simplifying assumption, we can decompose the total eigenvector of a many-electron molecule into separate contributions from individual electron stationary states (orbitals), each of which are obtained under the one-electron approximation. (Luckily, chemists and physicists can often (but not always) use this "single-electron approximation.") In this sense, in a many-electron system, an orbital can be considered as the stationary state of an individual electron in the system.

In chemistry, calculation of molecular orbitals typically also assume the Born–Oppenheimer approximation.

Relaxation to a Stationary State

Stationary and Non-stationary States

- Stationary state, by definition, means that the currents are steady and there is no net charge movement, i.e.

$$\nabla.\vec{\mathbf{J}}_s = 0 \text{ or } \frac{\partial \rho}{\partial t} = 0$$

These statements are equivalent, due to continuity.

- If the initial distribution of charges and currents does not satisfy the above criteria, they will redistribute themselves so that a stationary state is reached.
- This process of "relaxation" happens over a time scale that is characteristic of the medium, called the relaxation time.

Relaxation Time

- The continuity equation, combining with $\nabla.\vec{\mathbf{D}} = p$, gives

$$\nabla.\frac{\partial \vec{\mathbf{D}}}{\partial t} = -\nabla.\vec{\mathbf{J}}$$

- Using $\vec{\mathbf{D}} = \varepsilon \vec{\mathbf{E}}$ and $\vec{\mathbf{J}} = \sigma \vec{\mathbf{E}}$,

$$\nabla.(1 + \frac{\varepsilon}{\sigma}\frac{\partial}{\partial t})\vec{\mathbf{J}} = 0$$

- The solution to this differential equation is

$$\vec{\mathbf{J}} = \vec{\mathbf{J}}_s + (\vec{\mathbf{J}}_0 - \vec{\mathbf{J}}_s)e^{-t/\tau}$$

  where J_0 is the initial current distribution

- $\tau = \varepsilon / \sigma$ is the relaxation time
- $\frac{\partial \rho}{\partial t} = -\nabla.\vec{\mathbf{J}}, \vec{\mathbf{E}} = \vec{\mathbf{J}} / \sigma$, etc. relax at the same rate.

Time-dependent Electric Field

- No free charges, no external EMF sources. Maxwell $\Rightarrow$

$$\nabla\times(\nabla\times\vec{\mathbf{E}}) = -\frac{\partial}{\partial t}(\nabla\times\mu\vec{\mathbf{H}})$$

$$\nabla(\nabla.\vec{\mathbf{E}}) - \nabla^2\vec{\mathbf{E}} = -\mu\frac{\partial}{\partial t}(\vec{\mathbf{J}}_{\mathrm{fr}} + \varepsilon\frac{\partial\vec{\mathbf{E}}}{\partial t})$$

- This gives the second order partial differential equation

$$\nabla^2\vec{\mathbf{E}} - \mu\sigma\frac{\partial\vec{\mathbf{E}}}{\partial t} - \mu\varepsilon\frac{\partial^2\vec{\mathbf{E}}}{\partial t^2} = 0$$

- Depending on whether the $(\partial^2\vec{\mathbf{E}}/\partial t^2)$ term dominates or the $(\partial\vec{\mathbf{E}}/\partial t)$ one, we'll get two different extremes of behaviour. The former will lead to a propagating wave, the latter will lead to a diffusion equation, corresponding to a decaying wave.

Looking for solution of the form $\vec{\mathbf{E}}(\vec{\mathrm{x}})e^{-i\omega t}$

- The differential equation becomes

$$\nabla^2\vec{\mathbf{E}} + \mu\varepsilon\omega^2(1 + \frac{i\sigma}{\varepsilon\omega})\vec{\mathbf{E}} = 0$$

- There are two time scales here: $1/\omega$ and $\tau = \varepsilon/\sigma$

$$\nabla^2\vec{\mathbf{E}} + \mu\varepsilon\omega^2(1 + \frac{i}{\tau\omega})\vec{\mathbf{E}} = 0$$

- When $\tau\omega >> 1$

$$\nabla^2\vec{\mathbf{E}} + \mu\varepsilon\omega^2\vec{\mathbf{E}} = 0$$

which is a wave propagating with speed $\mathrm{c} = 1/\sqrt{\mu\varepsilon}$

- When $\tau\omega << 1$

$$\nabla^2\vec{\mathbf{E}} + \frac{i\omega}{\tau \mathrm{c}^2}\vec{\mathbf{E}} = 0$$

which is the equation for diffusion. In the context of EM waves, this will lead to a decaying solution.

Plane Wave

In the physics of wave propagation, a plane wave (also spelled planewave) is a wave whose wavefronts (surfaces of constant phase) are infinite parallel planes. Mathematically a plane wave takes the form

$$A(\vec{x},t)=f\left(\frac{\vec{n}}{c}\cdot\vec{x}-t\right)$$

in which the arbitrary (scalar or vector) function $\tau \mapsto f(\tau)$ gives the variation of the wave's amplitude, and the fixed unit vector $|\vec{n}|=1$ is the wave's direction of propagation. The solutions in $\vec{x}$ of

$$\tfrac{\vec{n}}{c}\cdot\vec{x}-t=\text{const.}$$

comprise the plane with normal vector $\vec{n}$. Thus, the points of equal field value of $A(\vec{x},t)$ always form a plane in space. This plane then shifts with time t, along the direction of propagation $\vec{n}$ with velocity c.

The term is often used to mean the special case of a monochromatic, homogeneous plane wave. A *monochromatic* (constant-frequency) plane wave is one in which the amplitude is a sinusoidal function of x and t. A *homogeneous* plane wave is one in which the planes of constant phase are perpendicular to the direction of propagation $\vec{n}$.

It is not possible in practice to have a true plane wave because it would have to fill all space and thus would require infinite energy; only a plane wave of infinite extent will propagate as a plane wave. However, many waves are approximately plane waves in a localized region of space. For example, a localized source such as an antenna produces a field that is approximately a plane wave far from the antenna in its far-field region. Similarly, if the length scales are much longer than the wave's wavelength, as is often the case for light in the field of optics, one can treat the waves as light rays which correspond locally to plane waves.

Mathematical Formalisms

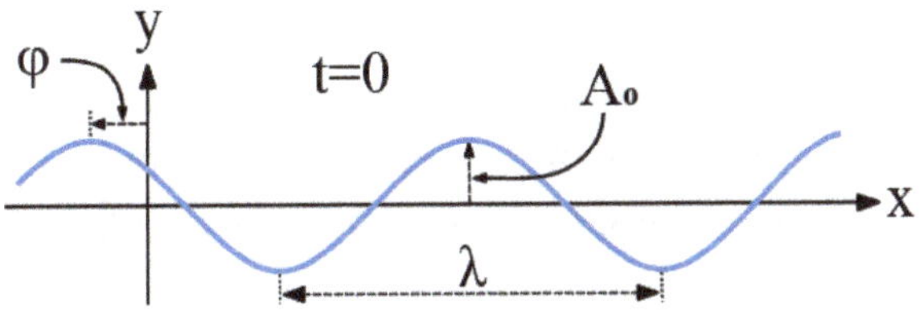

At time equals zero a positive phase shift results in the wave being shifted toward the left.

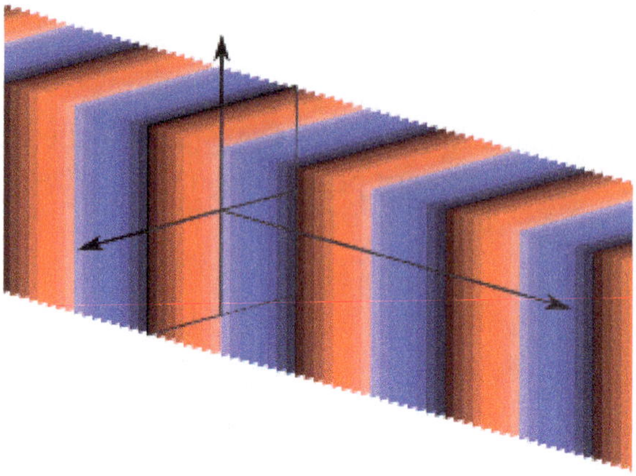

Image of a 3D plane wave. Each color represents a different phase of the wave.

Two functions that meet the above criteria of having a constant frequency (being harmonic) and constant amplitude are the sine and cosine functions. One of the simplest ways to use such a sinusoid involves defining it along the direction of the x-axis. The equation below, which is illustrated toward the right, uses the cosine function to represent a harmonic and homogeneous plane wave travelling in the positive x direction.

$$A(x,t) = A_o \cos(kx - \omega t + \varphi)$$

In the above equation:

- $A(x,t)$ is the magnitude or disturbance of the wave at a given point in space and time. An example would be to let $A(x,t)$ represent the variation of air pressure relative to the norm in the case of a sound wave.
- A_o is the amplitude of the wave which is the peak magnitude of the oscillation.
- k is the wave's wave number or more specifically the *angular* wave number and equals 2π/λ, where λ is the wavelength of the wave. k has the units of radians per unit distance and is a measure of how rapidly the disturbance changes over a given distance at a particular point in time.
- x is a point along the x-axis. y and z are not part of the equation because the wave's magnitude and phase are the same at every point on any given y-z plane. This equation defines what that magnitude and phase are.
- ω is the wave's angular frequency which equals 2π/T, where T is the period of the wave. ω has the units of radians per unit time and is a measure of how rapidly the disturbance changes over a given length of time at a particular point in space.
- t is a given point in time

- φ is the phase shift of the wave and has the units of radians. Note that a positive phase shift, at a given moment of time, shifts the wave in the negative x-axis direction. A phase shift of 2π radians shifts it exactly one wavelength.

Other formalisms which directly use the wave's wavelength λ, period T, frequency f and velocity c are below.

$$A = A_o \cos[2\pi(x/\lambda - t/T) + \varphi]$$
$$A = A_o \cos[2\pi(x/\lambda - ft) + \varphi]$$
$$A = A_o \cos[(2\pi/\lambda)(x - ct) + \varphi]$$

To appreciate the equivalence of the above set of equations note that $f = 1/T$ and $c = \lambda/T = \omega/k$

Arbitrary Direction

A more generalized form is used to describe a plane wave traveling in an arbitrary direction. It uses vectors in combination with the vector dot product.

$$A(\mathbf{r}, t) = A_o \cos(\mathbf{k} \cdot \mathbf{r} - \omega t + \varphi)$$

here:

- $\mathbf{k}$ is the wave vector which only differs from a wave number in that it has a direction as well as a magnitude. This means that, $|K| = k = 2\pi/\lambda$. The direction of the wave vector is ordinarily that direction which the plane wave is traveling, but it can differ slightly in an anisotropic medium.

- $\cdot$ is the vector dot product.

- $\mathbf{r}$ is the position vector which defines a point in three-dimensional space.

Complex Exponential Form

Many choose to use a more mathematically versatile formulation that utilizes the complex number plane. It requires the use of the natural exponent e and the imaginary number i.

$$U(\mathbf{r}, t) = A_o e^{i(\mathbf{k} \cdot \mathbf{r} - \omega t + \varphi)}$$

To appreciate this equation's relationship to the earlier ones, below is this same equation expressed using sines and cosines. Observe that the first term equals the real form of the plane wave just discussed.

$$U(\mathbf{r},t) = A_0 \cos(\mathbf{k}\cdot\mathbf{r} - \omega t + \varphi) + iA_0 \sin(\mathbf{k}\cdot\mathbf{r} - \omega t + \varphi)$$
$$U(\mathbf{r},t) = A(\mathbf{r},t) + iA_0 \sin(\mathbf{k}\cdot\mathbf{r} - \omega t + \varphi)$$

$$U(\mathbf{r},t) = A_0 e^{i(\mathbf{k}\cdot\mathbf{r} - \omega t + \varphi)}$$

equals

$$U(\mathbf{r},t) = A_0 e^{i(\mathbf{k}\cdot\mathbf{r} - \omega t)} e^{i\varphi}$$

one can absorb the phase factor $e^{i\varphi}$ into a complex amplitude by letting $U_0 = A_0 e^{i\varphi}$, resulting in the more compact equation

$$U(\mathbf{r},t) = U_0 e^{i(\mathbf{k}\cdot\mathbf{r} - \omega t)}$$

While the complex form has an imaginary component, after the necessary calculations are performed in the complex plane, its real value can be extracted giving a real valued equation representing an actual plane wave.

$$\mathrm{Re}[U(\mathbf{r},t)] = A(\mathbf{r},t) = A_0 \cos(\mathbf{k}\cdot\mathbf{r} - \omega t + \varphi)$$

The main reason one would choose to work with complex exponential form of plane waves is that complex exponentials are often algebraically easier to handle than the trigonometric sines and cosines. Specifically, the angle-addition rules are extremely simple for exponentials.

Additionally, when using Fourier analysis techniques for waves in a lossy medium, the resulting attenuation is easier to deal with using complex Fourier coefficients. It should be noted however that if a wave is traveling through a lossy medium, the amplitude of the wave is no longer constant, and therefore the wave is strictly speaking no longer a true plane wave.

In quantum mechanics the solutions of the Schrödinger wave equation are by their very nature complex and in the simplest instance take a form identical to the complex plane wave representation above. The imaginary component in that instance however has not been introduced for the purpose of mathematical expediency but is in fact an inherent part of the "wave".

In special relativity, one can utilize an even more compact expression by using four-vectors.

The four-position $\mathbf{R} = (ct, \mathbf{r})$

The four-wavevector $\mathbf{K} = \left(\frac{\omega}{c}, \mathbf{k}\right)$

The scalar product $\mathbf{K} \cdot \mathbf{R} = \omega t - \mathbf{k} \cdot \mathbf{r}$

Thus,

$$\mathrm{U}(\mathbf{r}, \mathrm{t}) = \mathrm{U}_o \mathrm{e}^{\mathrm{i}(\mathbf{k} \cdot \mathbf{r} - \omega \mathrm{t})}$$

becomes

$$\mathrm{U}(\mathbf{R}) = \mathrm{U}_o \mathrm{e}^{-\mathrm{i}(\mathbf{K} \cdot \mathbf{R})}$$

Applications

These waves are solutions for a scalar wave equation in a homogeneous medium. For vector wave equations, such as the ones describing electromagnetic radiation or waves in an elastic solid, the solution for a homogeneous medium is similar: the *scalar* amplitude A_o is replaced by a constant *vector* A_o. For example, in electromagnetism A_o is typically the vector for the electric field, magnetic field, or vector potential. A transverse wave is one in which the amplitude vector is orthogonal to k, which is the case for electromagnetic waves in an isotropic medium. By contrast, a longitudinal wave is one in which the amplitude vector is parallel to k, such as for acoustic waves in a gas or fluid.

The plane-wave equation works for arbitrary combinations of ω and k, but any real physical medium will only allow such waves to propagate for those combinations of ω and k that satisfy the dispersion relation of the medium. The dispersion relation is often expressed as a function, $\omega(\mathrm{k})$. The ratio $\omega/|\mathrm{k}|$ gives the magnitude of the phase velocity and $d\omega/d\mathrm{k}$ gives the group velocity. For electromagnetism in an isotropic medium with index of refraction n, the phase velocity is c/n, which equals the group velocity if the index is not frequency-dependent.

In linear uniform media, a wave solution can be expressed as a superposition of plane waves. This approach is known as the Angular spectrum method. The form of the planewave solution is actually a general consequence of translational symmetry. More generally, for periodic structures having discrete translational symmetry, the solutions take the form of Bloch waves, most famously in crystalline atomic materials but also in photonic crystals and other periodic wave equations. As another generalization, for structures that are only uniform along one direction x (such as a waveguide along the x direction), the solutions (waveguide modes) are of the form $\exp[i(kx-\omega t)]$

multiplied by some amplitude function $a(y,z)$. This is a special case of a separable partial differential equation.

Polarized Electromagnetic Plane Waves

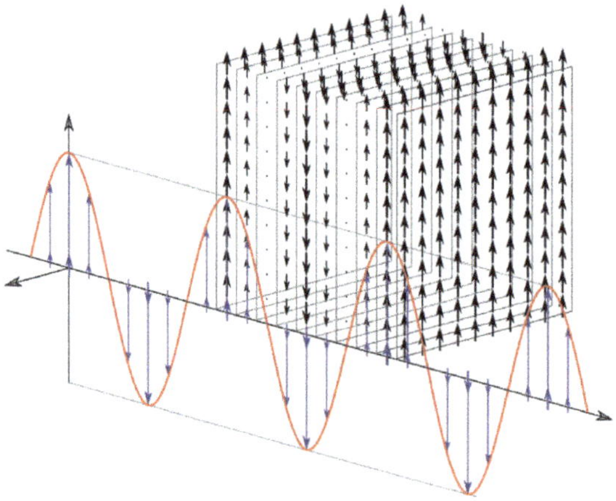

Linearly polarized light

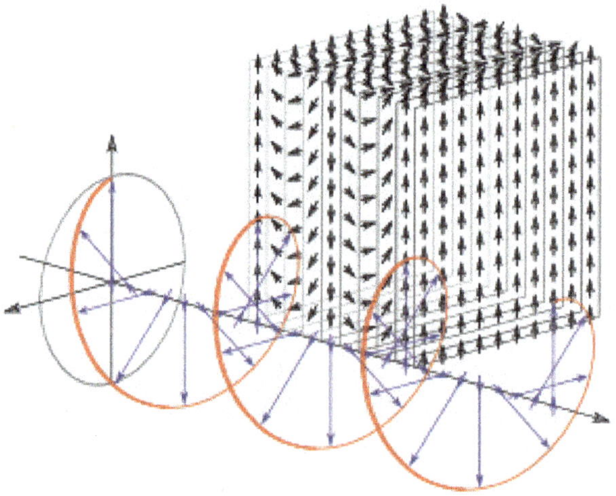

Circularly polarized light
The blocks of vectors represent how the magnitude and direction of the electric field is constant for an entire plane perpendicular to the direction of travel.

Represented in the first illustration toward the right is a linearly polarized, electromagnetic wave. Because this is a plane wave, each blue vector, indicating the perpendicular displacement from a point on the axis out to the sine wave, represents the magnitude and direction of the electric field for an entire plane that is perpendicular to the axis.

Represented in the second illustration is a circularly polarized, electromagnetic plane wave. Each blue vector indicating the perpendicular displacement from a point on the axis out to the helix, also represents the magnitude and direction of the electric field for an entire plane perpendicular to the axis.

In both illustrations, along the axes is a series of shorter blue vectors which are scaled down versions of the longer blue vectors. These shorter blue vectors are extrapolated out into the block of black vectors which fill a volume of space. Notice that for a given plane, the black vectors are identical, indicating that the magnitude and direction of the electric field is constant along that plane.

In the case of the linearly polarized light, the field strength from plane to plane varies from a maximum in one direction, down to zero, and then back up to a maximum in the opposite direction.

In the case of the circularly polarized light, the field strength remains constant from plane to plane but its direction steadily changes in a rotary type manner.

Not indicated in either illustration is the electric field's corresponding magnetic field which is proportional in strength to the electric field at each point in space but is at a right angle to it. Illustrations of the magnetic field vectors would be virtually identical to these except all the vectors would be rotated 90 degrees about the axis of propagation so that they were perpendicular to both the direction of propagation and the electric field vector.

The ratio of the amplitudes of the electric and magnetic field components of a plane wave in free space is known as the free-space wave-impedance, equal to 376.730313 ohms.

Propagating Plane Wave

Propagating (plane wave) solution for $\vec{\mathbf{E}}$

- $\omega\tau >> 1 \Rightarrow$ displacement current dominates over conduction current

$$\nabla^2\vec{\mathbf{E}} + \mu\varepsilon\omega^2\vec{\mathbf{E}} + i\omega\mu\sigma\vec{\mathbf{E}} = 0$$

- Plane wave: all fields are functions of the distance ζ of a plane from the origin. $\hat{n}$ is the normal to this plane.
- $\nabla \rightarrow \hat{n}(\partial / \partial\zeta)$
- Maxwell's equations in this language:

$$\hat{n}.\frac{\partial\vec{\mathbf{D}}}{\partial\zeta} = 0, \qquad \hat{n}\times\frac{\partial\vec{\mathbf{E}}}{\partial\zeta} = -\frac{\partial\vec{\mathbf{B}}}{\partial t}$$

$$\hat{n}.\frac{\partial\vec{\mathbf{B}}}{\partial\zeta} = 0, \qquad \hat{n}\times\frac{\partial\vec{\mathbf{H}}}{\partial\zeta} = \frac{\partial\vec{\mathbf{D}}}{\partial t}$$

Longitudinal Components of $\vec{E}$ and $\vec{B}$

$\vec{\mathbf{E}}_{\parallel}$: longitudinal component of $\vec{\mathbf{E}}$

- $(\partial\vec{\mathbf{D}}/\partial\zeta)$ equation and dot product of $\hat{n}$ with the $(\partial\vec{\mathbf{H}}/\partial\zeta)$ equation $\Rightarrow$

$$\frac{\partial\vec{\mathbf{E}}\,.\,\hat{n}}{\partial\zeta}=0\,,\;\frac{\partial\vec{\mathbf{E}}\,.\,\hat{n}}{\partial t}=0$$

- For non-conducting media (e.g. vacuum), $\vec{\mathbf{E}}_{\parallel}$ is a constant.

$\vec{\mathbf{B}}_{\parallel}$ **:** longitudinal component of $\vec{\mathbf{B}}$

- $(\partial\vec{\mathbf{B}}/\partial\zeta)$ equation and dot product of $\hat{n}$ with the $(\partial\vec{\mathbf{E}}/\partial\zeta)$ equation $\Rightarrow$

$$\frac{\partial\vec{\mathbf{B}}\,.\,\hat{n}}{\partial\zeta}=0\,,\;\frac{\partial\vec{\mathbf{B}}.\,\hat{n}}{\partial t}=0$$

- Only stationary longitudinal component of $\vec{\mathbf{B}}$ is possible, i.e. $\vec{\mathbf{B}}_{\parallel}$ is constant (note: we have taken $\mu=\mu_0$)

Transverse Components of $\vec{E}$ and $\vec{B}$

- Combining the two $\hat{n}\times$ equations:

$$\hat{n}\times\left(\frac{\partial^2\vec{\mathbf{E}}}{\partial\zeta^2}-\mu\varepsilon\frac{\partial^2\vec{\mathbf{E}}}{\partial t^2}\right)=0$$

- Differential equation for $\vec{\mathbf{E}}_{\perp}=\vec{\mathbf{E}}\times\hat{n}$
- General solution: $\vec{\mathbf{E}}_{\perp}=\vec{\mathbf{E}}_{\perp,0}\left[f(\zeta-ut)+g(\zeta+ut)\right]$
- If $\vec{\mathbf{E}}_{\perp}$ is sinusoidal:

$$\vec{\mathbf{E}}_{\perp}=\vec{\mathbf{E}}_{\perp,0}e^{-i(\omega t\pm k\zeta)}$$

- Direction of propagation $\vec{\mathbf{k}}\Rightarrow$

$$\vec{\mathbf{E}}_{\perp}=\vec{\mathbf{E}}_{\perp,0}e^{i(\vec{\mathbf{k}}.\vec{\mathbf{r}}-\omega t)}$$

- Using $\hat{n}\times(\partial\vec{\mathbf{E}}_{\perp}/\partial\zeta)=-\partial\vec{\mathbf{B}}_{\perp}/\partial t$,

$$i\vec{\mathbf{k}}\times\vec{\mathbf{E}}_{\perp}=i\omega\vec{\mathbf{B}}_{\perp}\Rightarrow\vec{\mathbf{B}}_{\perp}=\frac{\vec{\mathbf{k}}}{\omega}\times\vec{\mathbf{E}}$$

Propagating Wave in Short

- $\vec{\mathbf{E}}_{\parallel}$ and $\vec{\mathbf{B}}_{\parallel}$ are constants in space and time, hence not interesting for wave propagation
- $\vec{\mathbf{E}}_{\perp}$ and $\vec{\mathbf{B}}_{\perp}$ can have $e^{-\mathrm{Im}(k)x}$ dependence, with $\vec{\mathbf{B}}_{\perp} = (\vec{\mathbf{k}}/\omega)\times\vec{\mathbf{E}}$
- $\vec{\mathbf{E}}$ and $\vec{\mathbf{B}}$ fields are transverse to the direction of motion, and also orthogonal to each other.

Decaying Plane Wave

- When $\omega\tau << 1$, conduction current dominates over displacement current

$$\nabla^2\vec{\mathbf{E}} + \mu\varepsilon\omega^2\vec{\mathbf{E}} + \mathrm{i}\omega\mu\sigma\vec{\mathbf{E}} = 0$$

- The solution of the form $\vec{\mathbf{E}}_0 e^{\pm\mathrm{i}(\mathrm{kx}-\omega\mathrm{t})}$ implies

$$\mathrm{k}^2 = \frac{\mathrm{i}\omega}{\mathrm{C}^2\tau} = \frac{\omega}{\mathrm{C}^2\tau}\mathrm{e}^{i\pi/2}$$

$$\Rightarrow \mathrm{k} = \sqrt{\frac{\omega}{\mathrm{C}^2\tau}}\mathrm{e}^{i\pi/4} = \sqrt{\frac{\omega}{\mathrm{C}^2\tau}}\left(\frac{1+\mathrm{i}}{\sqrt{2}}\right)$$

- This gives

$$\vec{\mathbf{E}} = \vec{\mathbf{E}}_0 e^{\pm\mathrm{i}(\mathrm{Re}(\mathrm{k})\mathrm{x}-\omega\mathrm{t})}\ \mathrm{e}^{-\mathrm{Im}(\mathrm{k})\mathrm{x}}$$

- The wave then decays with a $\mathrm{e}^{-\mathrm{Im}(\mathrm{k})\mathrm{x}}$ dependence inside the conducting medium.

Skin Depth in Metals

- For metals, $\tau \sim 10^{-14}$ sec. So for $\omega < 10^{14}$, conduction current dominates.
- A wave incident on a metallic surface will decay as

$$\left|\vec{\mathbf{E}}\right| = \left|\vec{\mathbf{E}}_0\right|\mathrm{e}^{-\mathrm{r}/\delta}$$

where, from the last page, (check factor of 2)

$$\delta = \sqrt{\frac{2\mathrm{C}^2\tau}{\omega}} = \sqrt{\frac{2}{\sigma\omega}}$$

- Within a distance δ from the surface of the metal, the wave would decrease in magnitude by a factor 1/e. This δ is the "skin depth" of the metal. The surface currents will flow within this width.
- "Ideal" conductor $\Rightarrow \delta \to 0$.

Magnetostatics

Magnetostatics is the study of magnetic fields in systems where the currents are steady (not changing with time). It is the magnetic analogue of electrostatics, where the charges are stationary. The magnetization need not be static; the equations of magnetostatics can be used to predict fast magnetic switching events that occur on time scales of nanoseconds or less. Magnetostatics is even a good approximation when the currents are not static — as long as the currents do not alternate rapidly. Magnetostatics is widely used in applications of micromagnetics such as models of magnetic recording devices.

Applications

Magnetostatics as a Special Case of Maxwell's Equations

Starting from Maxwell's equations and assuming that charges are either fixed or move as a steady current $\mathbf{J}$, the equations separate into two equations for the electric field and two for the magnetic field. The fields are independent of time and each other. The magnetostatic equations, in both differential and integral forms, are shown in the table below.

Name	Partial differential form	Integral form
Gauss's law for magnetism:	$\nabla \cdot \mathbf{B} = 0$	$\oint_S \mathbf{B} \cdot \mathrm{d}\mathbf{S} = 0$
Ampère's law:	$\nabla \times \mathbf{H} = \mathbf{J}$	$\oint_C \mathbf{H} \cdot \mathrm{d}\mathbf{l} = I_{\mathrm{enc}}$

Where ∇ denotes divergence, and B is the magnetic flux density, the first integral is over a surface S with oriented surface element $d\mathbf{S}$. Where J is the current density and H is the magnetic field intensity, the second integral is a line integral around a closed loop C with line element $\mathbf{l}$. The current going through the loop is I_{enc}.

The quality of this approximation may be guessed by comparing the above equations with the full version of Maxwell's equations and considering the importance of the terms that have been removed. Of particular significance is the comparison of the $\mathbf{J}$ term against the $\partial \mathbf{D} / \partial t$ term. If the $\mathbf{J}$ term is substantially larger, then the smaller term may be ignored without significant loss of accuracy.

Re-introducing Faraday's Law

A common technique is to solve a series of magnetostatic problems at incremental time steps and then use these solutions to approximate the term $\partial \mathbf{B} / \partial t$. Plugging this result into Faraday's Law finds a value for $\mathbf{E}$ (which had previously been ignored). This method is not a true solution of Maxwell's equations but can provide a good approximation for slowly changing fields.

Solving for the Magnetic Field

Current Sources

If all currents in a system are known (*i.e.*, if a complete description of the current density $\mathbf{J}(\mathbf{r})$ is available) then the magnetic field can be determined, at a position r, from the currents by the Biot–Savart equation:

$$\mathbf{B}(\mathbf{r}) = \frac{\mu_0}{4\pi}\int \frac{\mathbf{J}(\mathbf{r}')\times(\mathbf{r}-\mathbf{r}')}{|\mathbf{r}-\mathbf{r}'|^3}\mathrm{d}^3\mathbf{r}'$$

This technique works well for problems where the medium is a vacuum or air or some similar material with a relative permeability of 1. This includes air-core inductors and air-core transformers. One advantage of this technique is that, if a coil has a complex geometry, it can be divided into sections and the integral evaluated for each section. Since this equation is primarily used to solve linear problems, the contributions can be added. For a very difficult geometry, numerical integration may be used.

For problems where the dominant magnetic material is a highly permeable magnetic core with relatively small air gaps, a magnetic circuit approach is useful. When the air gaps are large in comparison to the magnetic circuit length, fringing becomes significant and usually requires a finite element calculation. The finite element calculation uses a modified form of the magnetostatic equations above in order to calculate magnetic potential. The value of $\mathbf{B}$ can be found from the magnetic potential.

The magnetic field can be derived from the vector potential. Since the divergence of the magnetic flux density is always zero,

$$\mathbf{B} = \nabla\times\mathbf{A},$$

and the relation of the vector potential to current is:

$$\mathbf{A}(\mathbf{r}) = \frac{\mu_0}{4\pi}\int \frac{\mathbf{J}(\mathbf{r}')}{|\mathbf{r}-\mathbf{r}'|}\mathrm{d}^3\mathbf{r}'.$$

Magnetization

Strongly magnetic materials (*i.e.*, ferromagnetic, ferrimagnetic or paramagnetic) have a magnetization that is primarily due to electron spin. In such materials the magnetization must be explicitly included using the relation

$$\mathbf{B} = \mu_0(\mathbf{M} + \mathbf{H}).$$

Except in metals, electric currents can be ignored. Then Ampère's law is simply

$$\nabla \times \mathbf{H} = 0.$$

This has the general solution

$$\mathbf{H} = -\nabla \Phi_M,$$

where Φ_M is a scalar potential. Substituting this in Gauss's law gives

$$\nabla^2 \Phi_M = \nabla \cdot \mathbf{M}.$$

Thus, the divergence of the magnetization, $\nabla \cdot \mathbf{M}$, has a role analogous to the electric charge in electrostatics and is often referred to as an effective charge density ρ_M.

The vector potential method can also be employed with an effective current density

$$\mathbf{J_M} = \nabla \times \mathbf{M}.$$

Energy in Static Electric Field

Work done in Vacuum

- Work done in increasing charge density by $\delta\rho(\vec{x})$:

$$\delta W = \int \phi(\vec{x})\delta\rho(\vec{x})dV = \varepsilon_0 \int \phi(\vec{x})\delta(\nabla . \vec{E})dV$$

- Integrate by parts

$$\delta W = \varepsilon_0 \int \nabla . (\phi\delta\vec{E})dV - \varepsilon_0 \int (\nabla\phi) . (\delta\vec{E})dV$$
$$\delta W = \varepsilon_0 \int \vec{E} . \delta\vec{E}dV$$

Total work done

$$W = \frac{\varepsilon_0}{2} \int \overrightarrow{E^2} dV$$

Work done in a Dielectric

- Work done in increasing charge density by $\delta\rho(\vec{x})$:

$$\delta W = \int \phi(\vec{x})\delta p(\vec{x})dV = \int \phi(\vec{x})\delta(\nabla . \vec{D})dV$$

- Integrate by parts

$$\delta W = \int \nabla.(\phi\delta\vec{D})dV - \int (\nabla\phi).(\delta\vec{D})dV$$
$$\delta W = \int \vec{E}.\delta\vec{D}dV$$

Only for a linear dielectric $\vec{D} = \varepsilon\vec{E}$

$$W = \frac{1}{2}\int \vec{E}.\vec{D}dV$$

Matching with Energy of Collection of Discrete Charges

- For N charges q_i ,we expect that the energy is

$$U = \frac{\varepsilon_0}{2}\sum_n q_i\phi_i$$

where ϕ_i is the potential at the ith charge q_i due to the others. Let $\vec{E}_i$: electric field due to charge i

- Using $U = (\varepsilon_0/2)\int \vec{E^2}dV$ we get

$$U = \frac{\varepsilon_0}{2}\int \Big(\sum_i \vec{E_i}^2 + \sum_{i\neq j}\vec{E_i}.\vec{E_j}\Big)dV$$
$$= U_0 + \frac{\varepsilon_0}{2}\sum_i \int \vec{E_i}.(-\nabla\phi_i)dV$$

- The first term, U_0 does not depend on the distribution of charges, it is the "self-energy"
- The second term matches with what we expect.

Energy in Static Magnetic Field

Differences Compared to the Electric Field

- A steady current has to be maintained, for which energy should continue to be supplied by an external electric field E_{ext}.,
- When this current is increased by a small amount, a back-EMF E_{ind} is induced, against which work needs to be done. (This is the analog of work done against a repulsive force when bringing a charge from infinity.)
- This work done will be stored in the magnetic field produced. The rest of the energy supplied by the external EMF will be dissipated as heat in the conductor.

- Total current $\vec{J} = \sigma(\vec{E}_{ext} + \vec{E}_{ind}) \Rightarrow \vec{J}^2 = \sigma(\vec{E}_{ext}.\vec{J} + \vec{E}_{ind}.\vec{J})$
- Rate of energy input:

$$\frac{dU_{in}}{dt} = \vec{E}_{ext}.\vec{J}dV = \int \frac{\vec{J}^2}{\sigma} dV = \int \vec{E}_{ind}.\vec{J}dV$$

- The first term is rate of energy dissipation as heat, the second term is rate of storage of energy in the magnetic field

Energy Stored in Static Magnetic Field

- Rate of energy storage in magnetic field: $(\vec{E} = \vec{E}_{ind})$

$$\begin{aligned}\frac{dU_m}{dt} &= -\int \vec{E}.\vec{J}dV = -\int \vec{E}.(\nabla \times \vec{H})dV \\ &= -\int \vec{H}.(\nabla \times \vec{E})dV + \int \nabla.(\vec{E} \times \vec{H})dV \\ &= \int \vec{H}.(\partial \vec{B}/\partial t)dV\end{aligned}$$

- Incremental energy stored: $\delta U_m = \int \vec{H}.\delta\vec{B}dV$

Only if $\vec{H}$ is linear in $\vec{B}$

$$\begin{aligned}U_m &= \frac{1}{2}\int \vec{H}.\vec{B}dV = \frac{1}{2}\int \vec{H}.(\nabla \times \vec{A})dV \\ &= \frac{1}{2}\left[\int \vec{A}.(\nabla \times \vec{H})dV - \int \nabla.(\vec{H} \times \vec{A})dV\right] \\ &= \frac{1}{2}\int \vec{A}.\vec{J}dV\end{aligned}$$

A note about $\vec{E} \times \vec{H}$

- We neglected the term $\int \nabla.(\vec{E} \times \vec{H})dV = \oint (\vec{E} \times \vec{H}).d\vec{S}$
- This is indeed valid for static electromagnetic fields
- But for time-dependent fields, we shall see later that the leading terms in $\vec{E}$ and $\vec{H}$ go as 1/r, so over the surface of a sphere with large radius r, the integral actually will have a constant nonzero vale.
- This will be the energy radiated away at infinity due to the changing currents. $\vec{N} \equiv \vec{E} \times \vec{H}$ will be defined as the Poynting vector, which gives the rate of loss of energy through radiation.

Energy of EM Waves

Quadratic Quantities and Factors of 2

- In Electrodynamics, for convenience, we often use notation involving complex numbers (mainly exponentials), e.g.

$$\vec{E} = \vec{E}_0 e^{i(kx-\omega t)} \text{ , } \vec{B} = -i\vec{B}_0 e^{i(kx-\omega t)}$$

when we actually want to represent

$$\vec{E} = \vec{E}_0 \cos(kx-\omega t) = \mathrm{Re}(\vec{E}_0 e^{i(kx-\omega t)})$$
$$\vec{B} = \vec{B}_0 \sin(kx-\omega t) = \mathrm{Re}(i\vec{B}_0 e^{i(kx-\omega t)})$$

- While performing calculations in complex notation and taking the real part of the final answer works as long as we are dealing with quantities linear in ~E or ~B, one has to be careful while dealing with quadratic (or higher order) quantities.

- For example, in the complex notation above,

$$\left\langle \left|\vec{E}\right|^2 \right\rangle = \left\langle \left|\vec{E}^* . \vec{E}\right| \right\rangle = \left|\vec{E}_0\right|^2$$

while the actual answer should be (using real notation)

$$\left\langle \left|\vec{E}\right|^2 \right\rangle = \left|\vec{E}_0\right|^2 \left\langle \cos^2(kx-\omega t) \right\rangle = \frac{1}{2}\left|\vec{E}_0\right|^2$$

Energy Density Stored in EM Fields

- We have already seen that the energy stored in electric field is $U_e = (1/2)\varepsilon_0\left|\vec{E}\right|^2$ (we showed this result for a static field). When the electric field represents a propagating wave, then taking into account the "factor of 2" for averaged quadratic quantities, we get

$$\langle U_e \rangle = \frac{1}{4}\varepsilon_0\left|\vec{E}_0\right|^2$$

- The energy stored in magnetic field is $U_m = (1/2)\left|\vec{B}\right|^2/\mu_0$ (we showed it for a static magnetic field). For a propagating wave, $\left|\vec{B}\right| = \left|\zeta\omega\right|\left|\vec{E}\right|$. Including the "factor of 2", we get

$$\langle U_m \rangle = \frac{1}{4}\frac{\left|\vec{B}_0\right|^2}{\mu_0} = \frac{1}{4}\frac{\left|\right|^2}{\omega^2\mu_0}\left|\vec{E}_0\right|^2 = \frac{1}{4}\varepsilon_0\left|\vec{E}_0\right|^2$$

- For a plane EM wave, energy stored in electric and magnetic field is equal. The total energy of an EM wave is

$$\langle U \rangle = \langle U_e \rangle + \langle U_m \rangle = \frac{1}{2}\varepsilon_0 \left|\vec{E}_0\right|^2$$

Energy Transported by the EM Fields

- The rate of energy transport is given by the Poynting vector,

$$\vec{N} = \vec{E} \times \vec{H}$$

- The time-averaged value of this quantity is

$$\left\langle\left|\vec{N}\right|\right\rangle = \frac{1}{2}\left|\vec{E}_0\right| . \left|\vec{H}_0\right| = \frac{1}{2}\left|\vec{E}_0\right| \frac{||}{\omega\mu}\left|\vec{E}_0\right| = \frac{1}{2}\sqrt{\frac{\varepsilon_0}{\mu_0}}\left|\vec{E}_0\right|^2$$

- Compared with the rate of energy consumption in a conductor, $(1/2)\sigma\left|\vec{E}_0\right|^2$ the quantity $\sqrt{\varepsilon_0/\mu_0}$ is termed the conductance of vacuum
- Similarly, $\sqrt{\varepsilon/\mu}$ is the conductance of a medium through which an EM wave propagates

Dielectric

A dielectric material (dielectric for short) is an electrical insulator that can be polarized by an applied electric field. When a dielectric is placed in an electric field, electric charges do not flow through the material as they do in a conductor, but only slightly shift from their average equilibrium positions causing dielectric polarization. Because of dielectric polarization, positive charges are displaced toward the field and negative charges shift in the opposite direction. This creates an internal electric field that reduces the overall field within the dielectric itself. If a dielectric is composed of weakly bonded molecules, those molecules not only become polarized, but also reorient so that their symmetry axes align to the field.

The study of dielectric properties concerns storage and dissipation of electric and magnetic energy in materials. Dielectrics are important for explaining various phenomena in electronics, optics, solid-state physics, and cell biophysics.

Terminology

While the term *insulator* implies low electrical conduction, *dielectric* typically means materials with a high polarizability. The latter is expressed by a number called the relative permittivity (also known in older texts as dielectric constant). The term insulator is generally used to indicate electrical obstruction while the term dielectric is used to indicate the energy storing capacity of the material (by means of polarization). A common example of a dielectric is the electrically insulating material between the metallic plates of a capacitor. The polarization of the dielectric by the applied electric field increases the capacitor's surface charge for the given electric field strength.

The term "dielectric" was coined by William Whewell (from "dia-electric") in response to a request from Michael Faraday. A *perfect dielectric* is a material with zero electrical conductivity (cf. perfect conductor), thus exhibiting only a displacement current; therefore it stores and returns electrical energy as if it were an ideal capacitor.

Electric Susceptibility

The electric susceptibility χ_e of a dielectric material is a measure of how easily it polarizes in response to an electric field. This, in turn, determines the electric permittivity of the material and thus influences many other phenomena in that medium, from the capacitance of capacitors to the speed of light.

It is defined as the constant of proportionality (which may be a tensor) relating an electric field E to the induced dielectric polarization density P such that

$$\mathbf{P} = \varepsilon_0 \chi_e \mathbf{E},$$

where ε_0 is the electric permittivity of free space.

The susceptibility of a medium is related to its relative permittivity ε_r by

$$\chi_e = \varepsilon_r - 1.$$

So in the case of a vacuum,

$$\chi_e = 0.$$

The electric displacement D is related to the polarization density P by

$$\mathbf{D} = \varepsilon_0 \mathbf{E} + \mathbf{P} = \varepsilon_0 (1 + \chi_e) \mathbf{E} = \varepsilon_r \varepsilon_0 \mathbf{E}.$$

Dispersion and Causality

In general, a material cannot polarize instantaneously in response to an applied field. The more general formulation as a function of time is

$$\mathbf{P}(t) = \varepsilon_0 \int_{-\infty}^{t} \chi_e(t - t') \mathbf{E}(t') dt'.$$

That is, the polarization is a convolution of the electric field at previous times with time-dependent susceptibility given by $\chi_e(\Delta t)$. The upper limit of this integral can be extended to infinity as well if one defines $\chi_e(\Delta t) = 0$ for $\Delta t < 0$. An instantaneous response corresponds to Dirac delta function susceptibility $\chi_e(\Delta t) = \chi_e\delta(\Delta t)$.

It is more convenient in a linear system to take the Fourier transform and write this relationship as a function of frequency. Due to the convolution theorem, the integral becomes a simple product,

$$\mathbf{P}(\omega) = \varepsilon_0 \chi_e(\omega)\mathbf{E}(\omega).$$

Note the simple frequency dependence of the susceptibility, or equivalently the permittivity. The shape of the susceptibility with respect to frequency characterizes the dispersion properties of the material.

Moreover, the fact that the polarization can only depend on the electric field at previous times (i.e., $\chi_e(\Delta t) = 0$ for $\Delta t < 0$, a consequence of causality, imposes Kramers–Kronig constraints on the real and imaginary parts of the susceptibility $\chi_e(\omega)$.

Dielectric Polarization

Basic Atomic Model

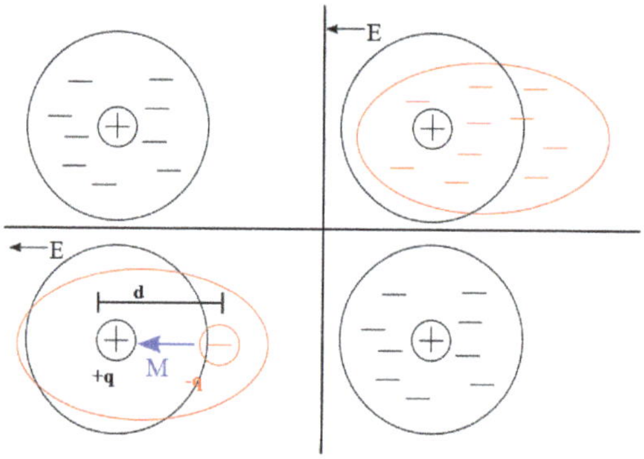

Electric field interaction with an atom under the classical dielectric model.

In the classical approach to the dielectric model, a material is made up of atoms. Each atom consists of a cloud of negative charge (electrons) bound to and surrounding a positive point charge at its center. In the presence of an electric field the charge cloud is distorted, as shown in the top right of the figure.

This can be reduced to a simple dipole using the superposition principle. A dipole is characterized by its dipole moment, a vector quantity shown in the figure as the blue arrow labeled *M*. It is the relationship between the electric field and the dipole moment that gives rise to the behavior of the dielectric. (Note that the dipole moment points in the same direction as the electric field in the figure. This isn't always the case, and is a major simplification, but is true for many materials.)

When the electric field is removed the atom returns to its original state. The time required to do so is the so-called relaxation time; an exponential decay.

This is the essence of the model in physics. The behavior of the dielectric now depends on the situation. The more complicated the situation, the richer the model must be to accurately describe the behavior. Important questions are:

- Is the electric field constant or does it vary with time? At what rate?
- Does the response depend on the direction of the applied field (isotropy of the material)?
- Is the response the same everywhere (homogeneity of the material)?
- Do any boundaries or interfaces have to be taken into account?
- Is the response linear with respect to the field, or are there nonlinearities?

The relationship between the electric field E and the dipole moment M gives rise to the behavior of the dielectric, which, for a given material, can be characterized by the function F defined by the equation:

$$\mathbf{M} = \mathbf{F}(\mathbf{E}).$$

When both the type of electric field and the type of material have been defined, one then chooses the simplest function *F* that correctly predicts the phenomena of interest. Examples of phenomena that can be so modeled include:

- Refractive index
- Group velocity dispersion
- Birefringence
- Self-focusing
- Harmonic generation

Dipolar Polarization

Dipolar polarization is a polarization that is either inherent to polar molecules (orientation polarization), or can be induced in any molecule in which the asymmetric distortion of the nuclei is possible (distortion polarization). Orientation polarization results from a permanent dipole, e.g., that arising from the 104.45° angle between the asymmetric bonds between oxygen and hydrogen atoms in the water molecule, which retains polarization in the absence of an external electric field. The assembly of these dipoles forms a macroscopic polarization.

When an external electric field is applied, the distance between charges within each permanent dipole, which is related to chemical bonding, remains constant in orientation polarization; however, the direction of polarization itself rotates. This rotation occurs on a timescale that depends on the torque and surrounding local viscosity of the molecules. Because the rotation is not instantaneous, dipolar polarizations lose the response to electric fields at the highest frequencies. A molecule rotates about 1 radian per picosecond in a fluid, thus this loss occurs at about 10^{11} Hz (in the microwave region). The delay of the response to the change of the electric field causes friction and heat.

When an external electric field is applied at infrared frequencies or less, the molecules are bent and stretched by the field and the molecular dipole moment changes. The molecular vibration frequency is roughly the inverse of the time it takes for the molecules to bend, and this distortion polarization disappears above the infrared.

Ionic Polarization

Ionic polarization is polarization caused by relative displacements between positive and negative ions in ionic crystals (for example, NaCl).

If a crystal or molecule consists of atoms of more than one kind, the distribution of charges around an atom in the crystal or molecule leans to positive or negative. As a result, when lattice vibrations or molecular vibrations induce relative displacements of the atoms, the centers of positive and negative charges are also displaced. The locations of these centers are affected by the symmetry of the displacements. When the centers don't correspond, polarizations arise in molecules or crystals. This polarization is called ionic polarization.

Ionic polarization causes the ferroelectric effect as well as dipolar polarization. The ferroelectric transition, which is caused by the lining up of the orientations of permanent dipoles along a particular direction, is called an order-disorder phase transition. The transition caused by ionic polarizations in crystals is called a displacive phase transition.

Ionic Polarization of Cells

Ionic polarization enables the production of energy-rich compounds in cells (the proton pump in mitochondria) and, at the plasma membrane, the establishment of the resting potential, energetically unfavourable transport of ions, and cell-to-cell communication (the Na+/K+-ATPase).

All cells in animal body tissues are electrically polarized – in other words, they maintain a voltage difference across the cell's plasma membrane, known as the membrane potential. This electrical polarization results from a complex interplay between protein structures embedded in the membrane called ion pumps and ion channels.

In neurons, the types of ion channels in the membrane usually vary across different parts of the cell, giving the dendrites, axon, and cell body different electrical properties.

As a result, some parts of the membrane of a neuron may be excitable (capable of generating action potentials), whereas others are not.

Dielectric Dispersion

In physics, dielectric dispersion is the dependence of the permittivity of a dielectric material on the frequency of an applied electric field. Because there is a lag between changes in polarization and changes in the electric field, the permittivity of the dielectric is a complicated function of frequency of the electric field. Dielectric dispersion is very important for the applications of dielectric materials and for the analysis of polarization systems.

This is one instance of a general phenomenon known as material dispersion: a frequency-dependent response of a medium for wave propagation.

When the frequency becomes higher:

1. dipolar polarization can no longer follow the oscillations of the electric field in the microwave region around 10^{10} Hz;
2. ionic polarization and molecular distortion polarization can no longer track the electric field past the infrared or far-infrared region around 10^{13} Hz, ;
3. electronic polarization loses its response in the ultraviolet region around 10^{15} Hz.

In the frequency region above ultraviolet, permittivity approaches the constant ε_0 in every substance, where ε_0 is the permittivity of the free space. Because permittivity indicates the strength of the relation between an electric field and polarization, if a polarization process loses its response, permittivity decreases.

Dielectric Relaxation

Dielectric relaxation is the momentary delay (or lag) in the dielectric constant of a material. This is usually caused by the delay in molecular polarization with respect to a changing electric field in a dielectric medium (e.g., inside capacitors or between two large conducting surfaces). Dielectric relaxation in changing electric fields could be considered analogous to hysteresis in changing magnetic fields (for inductors or transformers). Relaxation in general is a delay or lag in the response of a linear system, and therefore dielectric relaxation is measured relative to the expected linear steady state (equilibrium) dielectric values. The time lag between electrical field and polarization implies an irreversible degradation of Gibbs free energy.

In physics, dielectric relaxation refers to the relaxation response of a dielectric medium to an external, oscillating electric field. This relaxation is often described in terms of permittivity as a function of frequency, which can, for ideal systems, be described by the Debye equation. On the other hand, the distortion related to ionic and electronic polarization shows behavior of the resonance or oscillator type. The character of the distortion process depends on the structure, composition, and surroundings of the sample.

Debye Relaxation

Debye relaxation is the dielectric relaxation response of an ideal, noninteracting population of dipoles to an alternating external electric field. It is usually expressed in the complex permittivity ε of a medium as a function of the field's frequency ω:

$$\hat{\varepsilon}(\omega) = \varepsilon_\infty + \frac{\Delta\varepsilon}{1 + i\omega\tau},$$

where ε_∞ is the permittivity at the high frequency limit, $\Delta\varepsilon = \varepsilon_s - \varepsilon_\infty$ where ε_s is the static, low frequency permittivity, and τ is the characteristic relaxation time of the medium. Separating the real and imaginary parts of the complex dielectric permittivity yields:

$$\varepsilon' = \varepsilon_\infty + \frac{(\varepsilon_s - \varepsilon_\infty)}{(1 + \omega^2\tau^2)}$$

$$\varepsilon'' = \frac{(\varepsilon_s - \varepsilon_\infty)\omega\tau}{1 + \omega^2\tau^2}$$

The dielectric loss is also represented by:

$$\tan\delta = \frac{\varepsilon''}{\varepsilon'} = \frac{(\varepsilon_s - \varepsilon_\infty)\omega\tau}{\varepsilon_s + \varepsilon_\infty\omega^2\tau^2}$$

This relaxation model was introduced by and named after the physicist Peter Debye (1913). It is characteristic for dynamic polarization with only one relaxation time.

Variants of the Debye Equation

- Cole–Cole equation

This equation is used when the dielectric loss peak shows symmetric broadening

- Cole–Davidson equation

This equation is used when the dielectric loss peak shows asymmetric broadening

- Havriliak–Negami relaxation

This equation considers both symmetric and asymmetric broadening

- Kohlrausch–Williams–Watts function (Fourier transform of stretched exponential function)

Paraelectricity

Paraelectricity is the ability of many materials (specifically ceramics) to become polarized under an applied electric field. Unlike ferroelectricity, this can happen even if there is no permanent electric dipole that exists in the material, and removal of the fields results in the polarization in the material returning to zero. The mechanisms that cause paraelectric behaviour are the distortion of individual ions (displacement of the electron cloud from the nucleus) and polarization of molecules or combinations of ions or defects.

Paraelectricity can occur in crystal phases where electric dipoles are unaligned and thus have the potential to align in an external electric field and weaken it.

An example of a paraelectric material of high dielectric constant is strontium titanate.

The $LiNbO_3$ crystal is ferroelectric below 1430 K, and above this temperature it transforms into a disordered paraelectric phase. Similarly, other perovskites also exhibit paraelectricity at high temperatures.

Paraelectricity has been explored as a possible refrigeration mechanism; polarizing a paraelectric by applying an electric field under adiabatic process conditions raises the temperature, while removing the field lowers the temperature. A heat pump that operates by polarizing the paraelectric, allowing it to return to ambient temperature (by dissipating the extra heat), bringing it into contact with the object to be cooled, and finally depolarizing it, would result in refrigeration.

Tunability

Tunable dielectrics are insulators whose ability to store electrical charge changes when a voltage is applied.

Generally, strontium titanate ($SrTiO_3$) is used for devices operating at low temperatures, while barium strontium titanate ($Ba_{1-x}Sr_xTiO_3$) substitutes for room temperature devices. Other potential materials include microwave dielectrics and carbon nanotube (CNT) composites.

In 2013 multi-sheet layers of strontium titanate interleaved with single layers of strontium oxide produced a dielectric capable of operating at up to 125 GHz. The material was created via molecular beam epitaxy. The two have mismatched crystal spacing that produces strain within the strontium titanate layer that makes it less stable and tunable.

Systems such as $Ba_1{-}xSr_xTiO_3$ have a paraelectric–ferroelectric transition just below ambient temperature, providing high tunability. Such films suffer significant losses arising from defects.

Applications

Capacitors

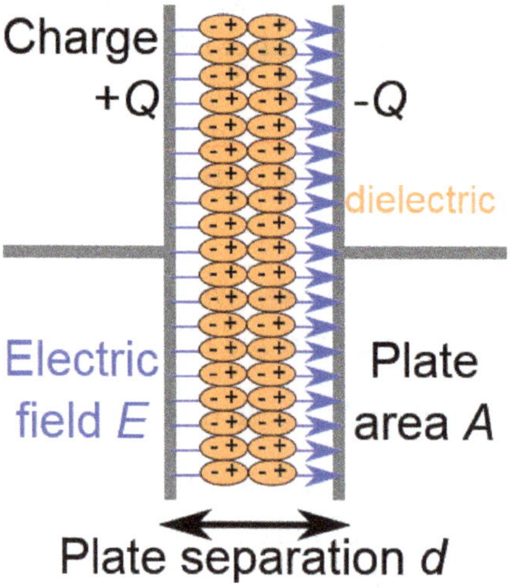

Charge separation in a parallel-plate capacitor causes an internal electric field.
A dielectric (orange) reduces the field and increases the capacitance.

Commercially manufactured capacitors typically use a solid dielectric material with high permittivity as the intervening medium between the stored positive and negative charges. This material is often referred to in technical contexts as the *capacitor dielectric.*

The most obvious advantage to using such a dielectric material is that it prevents the conducting plates, on which the charges are stored, from coming into direct electrical contact. More significantly, however, a high permittivity allows a greater stored charge at a given voltage. This can be seen by treating the case of a linear dielectric with permittivity ε and thickness d between two conducting plates with uniform charge density σ_ε. In this case the charge density is given by

$$\sigma_\varepsilon = \varepsilon \frac{V}{d}$$

and the capacitance per unit area by

$$c = \frac{\sigma_\varepsilon}{V} = \frac{\varepsilon}{d}$$

From this, it can easily be seen that a larger ε leads to greater charge stored and thus greater capacitance.

Dielectric materials used for capacitors are also chosen such that they are resistant to ionization. This allows the capacitor to operate at higher voltages before the insulating dielectric ionizes and begins to allow undesirable current.

Dielectric Resonator

A *dielectric resonator oscillator* (DRO) is an electronic component that exhibits resonance of the polarization response for a narrow range of frequencies, generally in the microwave band. It consists of a "puck" of ceramic that has a large dielectric constant and a low dissipation factor. Such resonators are often used to provide a frequency reference in an oscillator circuit. An unshielded dielectric resonator can be used as a dielectric resonator antenna (DRA).

Some Practical Dielectrics

Dielectric materials can be solids, liquids, or gases. In addition, a high vacuum can also be a useful, nearly lossless dielectric even though its relative dielectric constant is only unity.

Solid dielectrics are perhaps the most commonly used dielectrics in electrical engineering, and many solids are very good insulators. Some examples include porcelain, glass, and most plastics. Air, nitrogen and sulfur hexafluoride are the three most commonly used gaseous dielectrics.

- Industrial coatings such as parylene provide a dielectric barrier between the substrate and its environment.
- Mineral oil is used extensively inside electrical transformers as a fluid dielectric and to assist in cooling. Dielectric fluids with higher dielectric constants, such as electrical grade castor oil, are often used in high voltage capacitors to help prevent corona discharge and increase capacitance.
- Because dielectrics resist the flow of electricity, the surface of a dielectric may retain *stranded* excess electrical charges. This may occur accidentally when the dielectric is rubbed (the triboelectric effect). This can be useful, as in a Van de Graaff generator or electrophorus, or it can be potentially destructive as in the case of electrostatic discharge.
- Specially processed dielectrics, called electrets (which should not be confused with ferroelectrics), may retain excess internal charge or "frozen in" polarization. Electrets have a semipermanent electric field, and are the electrostatic equivalent to magnets. Electrets have numerous practical applications in the home and industry.
- Some dielectrics can generate a potential difference when subjected to mechanical stress, or (equivalently) change physical shape if an external voltage is applied across the material. This property is called piezoelectricity. Piezoelectric materials are another class of very useful dielectrics.
- Some ionic crystals and polymer dielectrics exhibit a spontaneous dipole

moment, which can be reversed by an externally applied electric field. This behavior is called the ferroelectric effect. These materials are analogous to the way ferromagnetic materials behave within an externally applied magnetic field. Ferroelectric materials often have very high dielectric constants, making them quite useful for capacitors.

EM Waves at Dielectric Boundaries: Reflection, Transmission

Reflection and Refraction

An EM wave is incident from one medium $(\varepsilon_1, \mu_1, n_1, c_1)$ to another medium $(\varepsilon_1, \mu_1, n_1, c_1)$, at an angle θ_1 with the normal to the boundary

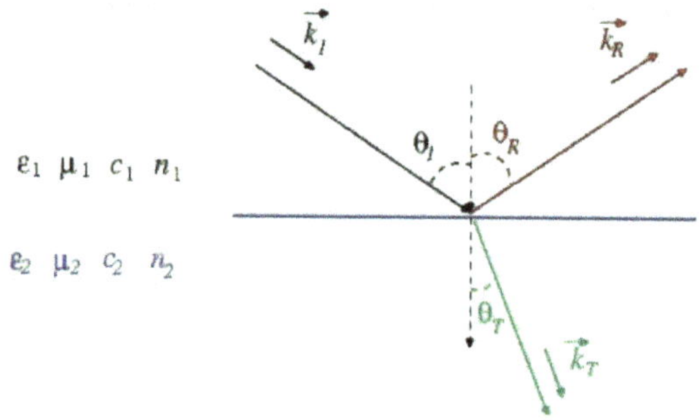

Incident, Reflected and Refracted Waves

Incident Wave

$$\vec{E}_I = \vec{E}_{I0} e^{i(\vec{k}_I \cdot \vec{r} - \omega t)}$$

$$\vec{B}_I = \frac{\vec{k}_I}{\omega} \times \vec{E}_I = \frac{1}{c_1}(\hat{k}_I \times \vec{E}_I)$$

Reflected Wave

$$\vec{E}_R = \vec{E}_{R0} e^{i(\vec{k}_R \cdot \vec{r} - \omega t)}$$

$$\vec{B}_R = \frac{\vec{k}_R}{\omega} \times \vec{E}_R = \frac{1}{c_1}(\hat{k}_R \times \vec{E}_R)$$

Transmitted Wave

$$\vec{E}_T = \vec{E}_{T0} e^{i(\vec{k}_T \cdot \vec{r} - \omega t)}$$

$$\vec{B}_T = \frac{\vec{k}_T}{\omega} \times \vec{E}_T = \frac{1}{c_2}(\hat{k}_T \times \vec{E}_T)$$

Boundary Conditions on Phases

$\vec{D}_\perp$ is continuous across the boundary

$$\varepsilon_1 \vec{E}_{I\perp} + \varepsilon_1 \vec{E}_{R\perp} = \varepsilon_2 \vec{E}_{T\perp}$$

$$\varepsilon_1 \vec{E}_{I\perp 0} e^{i(\vec{k}_I . \vec{r} - \omega t)} + \varepsilon_1 \vec{E}_{R\perp 0} e^{i(\vec{k}_R . \vec{r} - \omega t)} = \varepsilon_2 \vec{E}_{T\perp 0} e^{i(\vec{k}_T . \vec{r} - \omega t)}$$

- The equatity should be valid at all $\vec{r}$ on the boundary

$$\vec{k}_I \cdot \vec{r} = \vec{k}_R \cdot \vec{r} = \vec{k}_T \cdot \vec{r}$$

- With origin at the point of incidence:

$$\left|\vec{k}_I\right| r \sin\theta_I = \left|\vec{k}_R\right| r \sin\theta_R = \left|\vec{k}_T\right| r \sin\theta_T$$

- Using $|k_I| = |k_R|$ *and* $|k_T| / |k_I| = n_2 / n_1$,

$$\sin\theta_I = \sin\theta_R \;, \quad \frac{\sin\theta_I}{\sin\theta_T} = \frac{n_2}{n_1}$$

The first is the law of reflection the second is the Snell's law

Boundary Conditions on Amplitudes

Our discussion would have worked for any of the boundary conditions, we just took $\vec{D}_\perp$ as an example. Now we need not worry about the phases, since the laws of reflection and refraction derived there guarantee that the phase conditions will be satisfied.

Boundary Conditions

$$\varepsilon_1 \vec{E}_{I\perp 0} + \varepsilon_1 \vec{E}_{R\perp 0} = \varepsilon_2 \vec{E}_{T\perp 0}$$

$$\vec{B}_{I\perp 0} + \vec{B}_{R\perp 0} = \vec{B}_{T\perp 0}$$

$$\vec{E}_{I\|0} + \vec{E}_{R\|0} = \vec{E}_{T\|0}$$

$$\frac{1}{\mu_1}\vec{B}_{I\|0} + \frac{1}{\mu_1}\vec{B}_{R\|0} = \frac{1}{\mu_1}\vec{B}_{T\|0}$$

For convenience we'll divide the incident electric field into a component in the plane of incidence (the plane that contains $\vec{k}_I, \vec{k}_R, \vec{k}_T$) and a component normal to the plane of incidence. These two clearly won't interfere, and they can be added together at any time, using the principle of superposition, to get the net electric field.

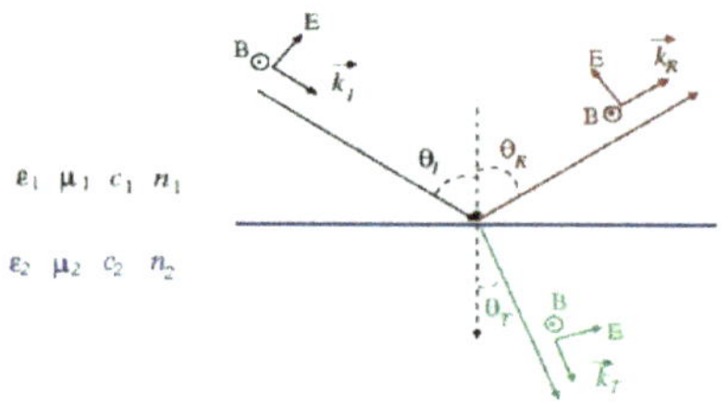

$\vec{E}$ in the plane of Incidence

Applying Boundary Conditions

Boundary conditions involving $\vec{E}$

$$-\varepsilon_1 E_{I0} \sin\theta_I + \varepsilon_1 E_{R0} \sin\theta_R = -\varepsilon_2 E_{T0} \sin\theta_T$$
$$-E_{I0} \cos\theta_I + E_{R0} \cos\theta_R = E_{T0} \cos\theta_T$$

Solution

$$E_{R0} = \left(\frac{\alpha - \beta}{\alpha + \beta}\right) E_{I0}\,,\ E_{T0} = \left(\frac{2}{\alpha + \beta}\right) E_{I0}$$

Where

$$\alpha \equiv \frac{\cos\theta_T}{\cos\theta_I}\,,\ \beta = \frac{\mu_1}{\mu_2}\frac{c_1}{c_2}$$

Boundary conditions involving $\vec{B}$ give exactly the same conditions.

Reflection and Transmission Coefficient

- Rate of energy transported by incoming wave normal to the boundary: (This is the same as the time-averaged magnitude of the Poynting vector $\vec{N} = \vec{E} \times \vec{H}$)

Incident wave: $I_I = \frac{1}{2}\varepsilon_1 c_1 \left|\vec{E}_{I0}\right|^2 \cos\theta_I$

Reflected wave: $I_R = \frac{1}{2}\varepsilon_1 c_1 \left|\vec{E}_{R0}\right|^2 \cos\theta_R$

Transmitted wave: $I_T = \frac{1}{2}\varepsilon_2 c_2 \left|\vec{E}_{T0}\right|^2 \cos\theta_T$

- Reflection coefficient

$$R = \frac{I_R}{I_I} = \left|\frac{\alpha - \beta}{\alpha + \beta}\right|^2$$

- Transmission coefficient

$$T = \frac{I_T}{I_I} = \frac{\varepsilon_2 c_2}{\varepsilon_1 c_1} \frac{\cos\theta_T}{\cos\theta_I} = \frac{4\,\mathrm{Re}(\alpha^* \beta)}{|\alpha+\beta|^2}$$

Comments on Reflection and Transmission Coefficients

- $R+T=1$ as expected
- $R=1, T=0$ possible if τ is purely imaginary.

$$\alpha = \frac{\sqrt{1-\sin^2\theta_T}}{\cos\theta_I} = \frac{\sqrt{1-(n_2/n_1)^2\sin^2\theta_T}}{\cos\theta_I},$$

so if sin $\theta_1 > (n_1/n_2)$ there is no transmission. This is the condition for Total Internal reflection.

- $R=0, T=1$ possible if $\alpha=\beta$,This condition takes a simple form if $\mu_1=\mu_2$ since then

$$\frac{\cos\theta_T}{\cos\theta_I} = \frac{c_1}{c_2} = \frac{\sin\theta_I}{\sin\theta_T} = \frac{c_1}{c_2}$$

This leads to sin $2\theta_I = \sin 2\theta_T$, that is $\theta_I + \theta_T = \pi/2$. In such a case, θ_I is called the Brewster's angle.

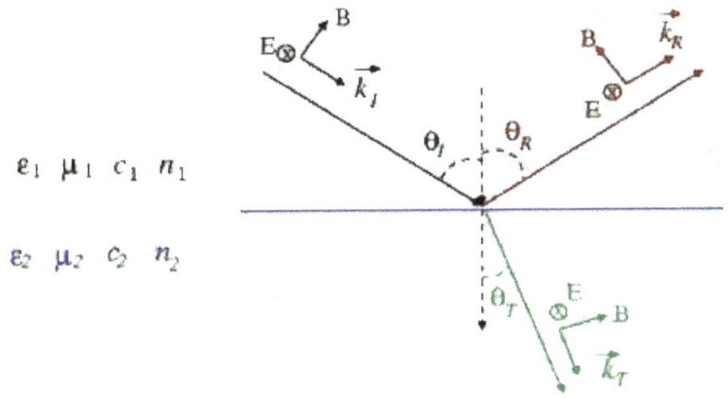

$\vec{E}$ normal to the plane of incidence

Comments on this Scenario

- The values for R and T will in general be different. In particular, R = 0 is not possible here.

- If an unpolarized wave is incident on a dielectric surface, the reflected and transmitted waves will therefore, in general, be polarized.

EM waves in Conductors: Inside and at the Boundary

Reflection from a Conducting Surface

- No wave is transmitted inside the conductor; i.e. fields inside the conductor are zero.
- For a normal incidence, $\vec{E}_I = -\vec{E}_R$ i.e. there is a phase-shift by π
- For incidence at an angle, the components of $\vec{E}_I$ and 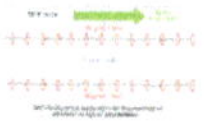 parallel to the boundary cancel, i.e. $\vec{E}_{I\parallel} = -\vec{E}_{R\parallel}$
- There will be charge oscillations at the metal surface corresponding to $\varepsilon_1(\vec{E}_{1\perp} + \vec{E}_{1\perp}) = \sigma_s$ where σ_s is the surface charge density
- The movements of these charges along the surface correspond to surface currents, which account for finite values of $\vec{H}_{I\parallel} + \vec{H}_{R\parallel}$ at the boundary.
- The net $\vec{B}$ normal to the surface vanishes, i.e. $\vec{B}_{I\perp} + \vec{B}_{R\perp} = 0$. This follows automatically from the $\vec{E}_{\parallel}$ conditions above.

Waveguide

A waveguide is a structure that guides waves, such as electromagnetic waves or sound, with minimal loss of energy by restricting expansion to one dimension or two. This is a similar effect to waves of water constrained within a canal, or why guns have barrels that restrict hot gas expansion to maximize energy transfer to their bullets. Without the physical constraint of a waveguide, waves are decreasing according to the inverse square law as they expand into three dimensional space.

There are different types of waveguides for each type of wave. The original and most common meaning is a hollow conductive metal pipe used to carry high frequency radio waves, particularly microwaves.

The geometry of a waveguide reflects its function. Slab waveguides confine energy in one dimension, fiber or channel waveguides in two dimensions. The frequency of the transmitted wave also dictates the shape of a waveguide: an optical fiber guiding high-frequency light will not guide microwaves of a much lower frequency. As a rule of thumb, the width of a waveguide needs to be of the same order of magnitude as the wavelength of the guided wave.

Some naturally occurring structures can also act as waveguides. The SOFAR channel layer in the ocean can guide the sound of whale song across enormous distances.

Principle of Operation

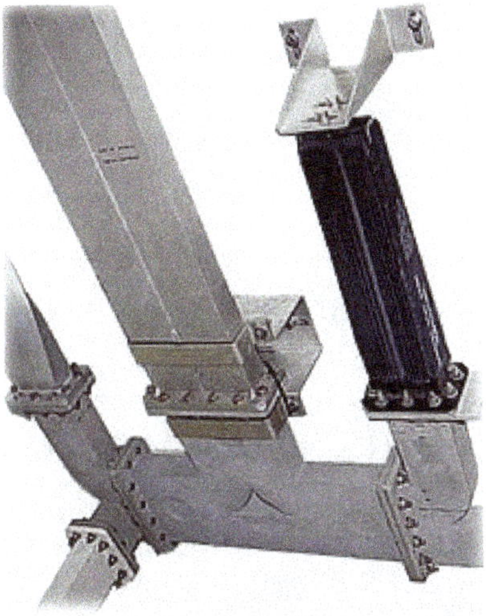

Example of waveguides and a diplexer in an air traffic control radar

Waves propagate in all directions in open space as spherical waves. The power of the wave falls with the distance R from the source as the square of the distance (inverse square law). A waveguide confines the wave to propagate in one dimension, so that, under ideal conditions, the wave loses no power while propagating. Due to total reflection at the walls, waves are confined to the interior of a waveguide.

History

The first structure for guiding waves was proposed by J. J. Thomson in 1893, and was first experimentally tested by Oliver Lodge in 1894. The first mathematical analysis of electromagnetic waves in a metal cylinder was performed by Lord Rayleigh in 1897. For sound waves, Lord Rayleigh published a full mathematical analysis of propagation modes in his seminal work, "The Theory of Sound".

The study of dielectric waveguides began as early as the 1920s, by several people, most famous of which are Rayleigh, Sommerfeld and Debye. Optical fiber began to receive special attention in the 1960s due to its importance to the communications industry.

The development of radio communication initially occurred at the lower frequencies because these could be more easily propagated over large distances. The long wavelengths made these frequencies unsuitable for use in hollow metal waveguides because of the impractically large diameter tubes required. Consequently, research into hollow metal waveguides stalled and the work of Lord Rayleigh was forgotten for a time and had to be rediscovered by others. Practical investigations resumed in the 1930s by George C. Southworth at Bell Labs and Wilmer L. Barrow at MIT. Southworth at first took the theory from papers on waves in dielectric rods because the work of Lord Rayleigh was unknown to him. This misled him somewhat; some of his experiments failed because he was not aware of the phenomenon of waveguide cutoff frequency found in Lord Rayleigh's work. Serious theoretical work was taken up by John R. Carson and Sallie P. Mead. This work led to the discovery that for the TE_{01} mode in circular waveguide losses go down with frequency and at one time this was a serious contender for the format for long distance telecommunications.

The importance of radar in World War II gave a great impetus to waveguide research, at least on the Allied side. The magnetron developed in 1940 by John Randall and Harry Boot at the University of Birmingham in the United Kingdom provided a good power source and made microwave radars feasible. The most important centre of research was at the Radiation Laboratory (Rad Lab) at MIT but many others took part in the US, and in the UK such as the Telecommunications Research Establishment. The head of the Fundamental Development Group at Rad Lab was Edward Mills Purcell. His researchers included Julian Schwinger, Nathan Marcuvitz, Carol Gray Montgomery, and Robert H. Dicke. Much of the Rad Lab work concentrated on finding lumped element models of waveguide structures so that components in waveguide could be analysed with standard circuit theory. Hans Bethe was also briefly at Rad Lab, but while there he produced his small aperture theory which proved important for waveguide cavity filters, first developed at Rad Lab. The German side, on the other hand, largely ignored the potential of waveguides in radar until very late in the war. So much so that when radar parts from a downed British plane were sent to Siemens & Halske for analysis, even though they were recognised as microwave components, their purpose could not be identified.

At that time, microwave techniques were badly neglected in Germany. It was generally believed that it was of no use for electronic warfare, and those who wanted to do research work in this field were not allowed to do so.

—H. Mayer, wartime vice-president of Siemens & Halske

German academics were even allowed to continue publicly publishing their research in this field because it was not felt to be important.

Immediately after World War II waveguide was the technology of choice in the microwave field. However, it has some problems; it is bulky, expensive to produce, and the cutoff frequency effect makes it difficult to produce wideband devices. Ridged waveguide can increase bandwidth beyond an octave, but a better solution is to use a technology working in TEM mode (that is, non-waveguide) such as coaxial conductors since TEM does not have a cutoff frequency. A shielded rectangular conductor can also be used and this which certain manufacturing advantages over coax and can be seen as the forerunner of the planar technologies (stripline and microstrip). However, planar technologies really started to take off when printed circuits were introduced. These methods are significantly cheaper than waveguide and have largely taken its place in most bands. However, waveguide is still favoured in the higher microwave bands from around Ku band upwards.

Uses

Waveguide supplying power for the Argonne National Laboratory Advanced Photon Source.

The uses of waveguides for transmitting signals were known even before the term was coined. The phenomenon of sound waves guided through a taut wire have been known for a long time, as well as sound through a hollow pipe such as a cave or medical stethoscope. Other uses of waveguides are in transmitting power between the components of a system such as radio, radar or optical devices. Waveguides are the fundamental principle of guided wave testing (GWT), one of the many methods of non-destructive evaluation.

Specific Examples:

- Optical fibers transmit light and signals for long distances with low attenuation and a wide usable range of wavelengths.
- In a microwave oven a waveguide transfers power from the magnetron, where waves are formed, to the cooking chamber.
- In a radar, a waveguide transfers radio frequency energy to and from the antenna, where the impedance needs to be matched for efficient power transmission (see below).
- Rectangular and Circular waveguides are commonly used to connect feeds of parabolic dishes to their electronics, either low-noise receivers or power amplifier/transmitters.
- Waveguides are used in scientific instruments to measure optical, acoustic and elastic properties of materials and objects. The waveguide can be put in contact with the specimen (as in a medical ultrasonography), in which case the waveguide ensures that the power of the testing wave is conserved, or the specimen

may be put inside the waveguide (as in a dielectric constant measurement), so that smaller objects can be tested and the accuracy is better.

- Transmission lines are a specific type of waveguide, very commonly used.

Propagation Modes and Cutoff Frequencies

A propagation mode in a waveguide is one solution of the wave equations, or, in other words, the form of the wave. Due to the constraints of the boundary conditions, there are only limited frequencies and forms for the wave function which can propagate in the waveguide. The lowest frequency in which a certain mode can propagate is the cutoff frequency of that mode. The mode with the lowest cutoff frequency is the fundamental mode of the waveguide, and its cutoff frequency is the waveguide cutoff frequency.

Propagation modes are computed by solving the Helmholtz equation alongside a set of boundary conditions depending on the geometrical shape and materials bounding the region. The usual assumption for infinitely long uniform waveguides allows to assume a propagating form for the wave, i.e. stating that every field component has a known dependency on the propagation direction (i.e. z). More specifically, the common approach is to first replace all unknown time-varying unknown fields $u(x,y,z,t)$ (assuming for simplicity to describe the fields in cartesian components) with their complex phasors representation $U(x,y,z)$, sufficient to fully describe any infinitely long single-tone signal at frequency f, (angular frequency $\omega = 2\pi f$), and rewrite the Helmholtz equation and boundary conditions accordingly. Then, every unknown field is forced to have a form like $U(x,y,z) = \hat{U}(x,y)e^{-\gamma z}$, where the γ term represents the propagation constant (still unknown) along the direction along which the waveguide extends to infinity. The Helmholtz equation can be rewritten to accommodate such form and the resulting equality needs to be solved for γ and $\hat{U}(x,y)$, yielding in the end an eigenvalue equation for γ and a corresponding eigenfunction $\hat{U}(x,y)_\gamma$ for each solution of the former.

The propagation constant γ of the guided wave is complex, in general. For a lossless case, the propagation constant might be found to take on either real or imaginary values, depending on the chosen solution of the eigenvalue equation and on the angular frequency ω. When γ is purely real, the mode is said to be "below cutoff", since the amplitude of the field phasors tends to exponentially decrease with propagation; an imaginary γ, instead, represents modes said to be "in propagation" or "above cutoff", as the complex amplitude of the phasors does not change with z .

Impedance Matching

In circuit theory, the impedance is a generalization of electrical resistivity in the case of alternating current, and is measured in ohms (Ω). A waveguide in circuit theory is described by a transmission line having a length and characteristic impedance. In other words, the impedance indicates the ratio of voltage to current of the circuit component (in this case a waveguide) during propagation of the wave. This description of the waveguide was originally intended for alternating current, but is also suitable for electromagnetic and sound waves, once the wave and material properties (such as pressure, density, dielectric constant) are properly converted into electrical terms (current and impedance for example).

Impedance matching is important when components of an electric circuit are connected (waveguide to antenna for example): The impedance ratio determines how much of the wave is transmitted forward and how much is reflected. In connecting a waveguide to an antenna a complete transmission is usually required, so an effort is made to match their impedances.

The reflection coefficient can be calculated using: $\Gamma = \frac{Z_2 / Z_1 - 1}{Z_2 / Z_1 + 1}$, where Γ is the reflection coefficient (0 denotes full transmission, 1 full reflection, and 0.5 is a reflection of half the incoming voltage), Z_1 and Z_2 are the impedance of the first component (from which the wave enters) and the second component, respectively.

An impedance mismatch creates a reflected wave, which added to the incoming waves creates a standing wave. An impedance mismatch can be also quantified with the standing wave ratio (SWR or VSWR for voltage), which is connected to the impedance ratio and reflection coefficient by: $VSWR = \frac{|V|_{max}}{|V|_{min}} = \frac{1+|\Gamma|}{1-|\Gamma|}$, where $|V|_{min,max}$ are the minimum and maximum values of the voltage absolute value, and the VSWR is the voltage standing wave ratio, which value of 1 denotes full transmission, without reflection and thus no standing wave, while very large values mean high reflection and standing wave pattern.

Electromagnetic Waveguides

In this military radar, microwave radiation is transmitted between the source and the reflector by a waveguide. The figure suggests that microwaves leave the box in a circularly symmetric mode (allowing the antenna to rotate), then they are converted to a linear mode, and pass through a flexible stage. Their polarisation is then rotated in a twisted stage and finally they irradiate the parabolic antenna.

Waveguides can be constructed to carry waves over a wide portion of the electromagnetic spectrum, but are especially useful in the microwave and optical frequency ranges. Depending on the frequency, they can be constructed from either conductive or dielectric materials. Waveguides are used for transferring both power and communication signals.

Optical Waveguides

Waveguides used at optical frequencies are typically dielectric waveguides, structures in which a dielectric material with high permittivity, and thus high index of refraction, is surrounded by a material with lower permittivity. The structure guides optical waves by total internal reflection. An example of an optical waveguide is optical fiber.

Other types of optical waveguide are also used, including photonic-crystal fiber, which guides waves by any of several distinct mechanisms. Guides in the form of a hollow tube with a highly reflective inner surface have also been used as light pipes for illumination applications. The inner surfaces may be polished metal, or may be covered with a multilayer film that guides light by Bragg reflection (this is a special case of a photonic-crystal fiber). One can also use small prisms around the pipe which reflect light via total internal reflection —such confinement is necessarily imperfect, however, since total internal reflection can never truly guide light within a *lower*-index core (in the prism case, some light leaks out at the prism corners).

Acoustic Waveguides

An *acoustic waveguide* is a physical structure for guiding sound waves. A duct for sound propagation also behaves like a transmission line. The duct contains some medium, such as air, that supports sound propagation. Acoustic waveguides have no cutoff frequency, which is in sharp contrast to electromagnetic waveguides.

Mathematical Waveguides

Waveguides are interesting objects of study from a strictly mathematical perspective. A waveguide (or tube) is defined as type of boundary condition on the wave equation such that the wave function must be equal to zero on the boundary and that the allowed region is finite in all dimensions but one (an infinitely long cylinder is an example.) A large number of interesting results can be proven from these general conditions. It turns out that any tube with a bulge (where the width of the tube increases) admits at least one bound state. This can be shown using the variational principles. An interesting result by Jeffrey Goldstone and Robert Jaffe is that any tube of constant width with a twist, admits a bound state.

Sound Synthesis

Sound synthesis uses digital delay lines as computational elements to simulate wave propagation in tubes of wind instruments and the vibrating strings of string instruments.

Travelling Waves with the Same (x, y) Profile

- We are looking for waves travelling in z direction, while keeping the same (x, y) profile. I.e. the form

$$\vec{\mathbf{E}} = \vec{\mathbf{E}}^{0}(x,y)e^{i(k_z z-\omega t)}\ ,\ \vec{\mathbf{B}} = \vec{\mathbf{B}}^{0}(x,y)e^{i(k_z z-\omega t)}$$

- Maxwell's $(\nabla\times\vec{\mathbf{E}})$ and $(\nabla\times\vec{\mathbf{B}})$ equations then become

$$\frac{\partial E_y}{\partial x}-\frac{\partial E_x}{\partial y}=i\omega B_z\ ,\quad \frac{\partial B_y}{\partial x}-\frac{\partial B_x}{\partial y}=-\frac{i\omega}{c^2}E_z$$

$$\frac{\partial E_z}{\partial y}-ik_zE_y=i\omega B_x\ ,\quad \frac{\partial B_z}{\partial y}-ik_zB_y=-\frac{i\omega}{c^2}E_x$$

$$ik_zE_x-\frac{\partial E_z}{\partial x}=i\omega B_y\ ,\ ik_zB_x-\frac{\partial B_z}{\partial x}=-\frac{i\omega}{c^2}E_y$$

- Note that one can factor out the $e^{i(k_z z-\omega t)}$ dependence of E_x, E_y, E_z and B_x, B_y, B_z so now onwards they have no z- or t-dependence in this section.

- Using the last two lines (4 equations), one can write E_x, E_y, B_x, B_y in terms of the other two quantities, and B_z

All components in terms of E_z and B_z

$$E_x=\frac{1}{(\omega/c)^2-k_z^2}\left(k_z\frac{\partial E_z}{\partial x}+\omega\frac{\partial B_z}{\partial y}\right)$$

$$E_y=\frac{1}{(\omega/c)^2-k_z^2}\left(k_z\frac{\partial E_z}{\partial y}+\omega\frac{\partial B_z}{\partial x}\right)$$

$$B_x=\frac{i}{(\omega/c)^2-k_z^2}\left(k_z\frac{\partial B_z}{\partial x}+\frac{\omega}{c^2}\frac{\partial E_z}{\partial y}\right)$$

$$B_y=\frac{i}{(\omega/c)^2-k_z^2}\left(k_z\frac{\partial B_z}{\partial y}+\frac{\omega}{c^2}\frac{\partial E_z}{\partial x}\right)$$

- Note that if E_z and B_z both vanish (or are constants), no other components of $\vec{\mathbf{E}}$ or $\vec{\mathbf{B}}$ can survive (unless $\mathrm{k}_z = 0$ which case needs to be treated separately.)

- However E_z and B_z are not free parameters; the above equations just give four constraints on $\vec{\mathbf{E}}$ and $\vec{\mathbf{B}}$, two more constraints from the last page are still remaining.

Constraining E_z, B_z themselves

E_z, B_z themselves must satisfy consistency conditions

$$\frac{\partial E_y}{\partial x} - \frac{\partial E_x}{\partial y} = i\omega B_z$$

$$\frac{\partial B_y}{\partial x} - \frac{\partial B_x}{\partial y} = -\frac{i\omega}{c^2} E_z$$

These Correspond to

$$\left(\frac{\partial^2}{\partial x^2} + \frac{\partial^2}{\partial y^2} - k_z^2 + \frac{\omega^2}{c^2} \right) B_z = 0$$

$$\left(\frac{\partial^2}{\partial x^2} + \frac{\partial^2}{\partial y^2} - k_z^2 + \frac{\omega^2}{c^2} \right) E_z = 0$$

If there were no boundary conditions in the x- y plane, this would have a plane wave solution – a flat x - y profile. But conducting boundaries imply that these fields must have a non-trivial x-y profile.

EM wave Propagation in Waveguides

- Let us consider rectangular / circular hollow conducting cylinders, through which an EM wave will be "guided" by bending the boundaries of the cylinders.

- A simple solution would have been a plane wave travelling along z direction, such that $\vec{\mathbf{E}}$ and $\vec{\mathbf{B}}$ fields are transverse, $E_z = B_z = 0$. Such a solution is called as TEM (transverse electric and magnetic) mode.

- Such a mode is not possible in a hollow cylinder, proof given on the next page.

- However E_z and B_z can individually vanish, such modes are termed TE (Transverse electric: $E_z = 0$) and TM (Transverse magnetic: $B_z = 0$).

Hollow cylinder cannot have both $E_z = 0$ and $B_z = 0$

- Since $B_z = 0$ we have $\left(\nabla \times \vec{\mathbf{E}}\right)_z = -\partial B_z / \partial t = 0$. Then

$$\frac{\partial E_y}{\partial x} - \frac{\partial E_x}{\partial y} = 0$$

- Since E_x and E_y are independent of z, and $E_z = 0$ we get $\nabla \times \vec{\ } =$, i.e. $\vec{\mathbf{E}}$ can be written as $\vec{\mathbf{E}} = -\nabla\Phi$.

- In addition, no charges inside the cylinder, so $\nabla . E = 0$. That is, $\nabla^2 \Phi = 0$.

- Now we have a boundary value problem, with $\nabla^2\Phi = 0$ inside the boundary and Φ =constant on the complete boundary (the hollow conductor).
- This boundary value problem has a solution, Φ =constant everywhere, and the uniqueness theorem states that this is the only solution.
- Thus, there can be no electric / magnetic fields inside the waveguide.

TE and TM Modes

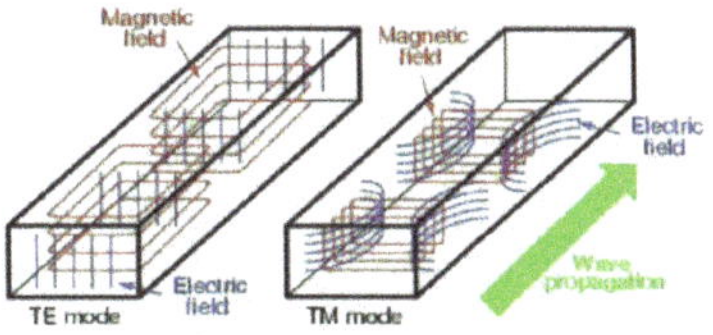

Magnetic flux lines appear as continuous loops
Electric flux lines appear with beginning and end points

TEM Mode

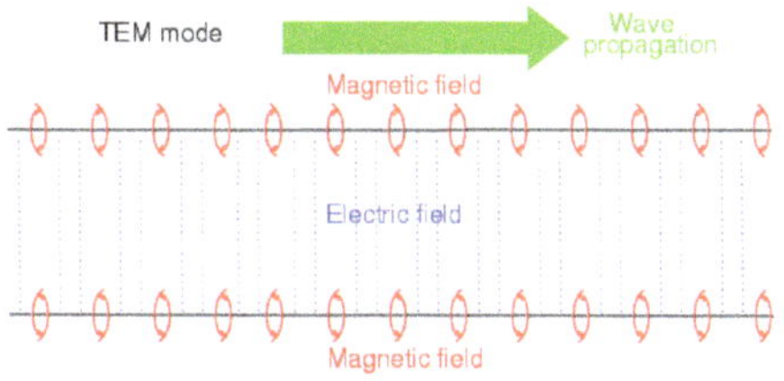

Both field planes perpendicular (transverse) to direction of signal propagation.

Rectangular Waveguide

TE modes ($E_z = 0$, $B_z \neq 0$) in a rectangular waveguide

- Let the walls of the waveguide be at y = 0; b and x = 0; a.

  The boundary conditions are then

  $E_x = 0 \; at \; y = 0, b$ and $E_y = 0 \; at \; x = 0, a$

- The equations that give $E_{x,y}$ in terms of B_z then imply $\frac{\partial B_z}{\partial y} = 0 \; at \; y = 0, b$ and $\frac{\partial B_z}{\partial y} = 0 \; at \; x = 0, a$
- The solution to the differential equation for B_z with these boundary conditions, is

$$B_z = A\cos(k_x x)\cos(k_y y)$$

where $k_x = (m\pi/a)$ and $k_y = (n\pi/b)$.

- Such a mode is called TE_{mn}. Note that at least one of m or n has to be nonzero, else all fields will vanish.

Cutoff Frequencies for TE Modes

- The TE_{mn} solution, when substituted in the differential equation for B_z, gives

$$-\left(\frac{m\pi}{a}\right)^2 - \left(\frac{n\pi}{b}\right)^2 - k_z^2 + (\omega/c)^2 = 0$$

- For consistency with the physical situation, k_z must be real; i.e. $k_z^2 > 0$ This gives the condition

$$\omega > c\sqrt{\left(\frac{m\pi}{a}\right)^2 + \left(\frac{n\pi}{b}\right)^2} \equiv \omega_{mn}$$

- Thus, for a TE mode TE_{mn} to propagate, it must have a minimum frequency ω_{mn}. A waveguide thus acts like a high-pass filter.

TM Modes

- A similar analysis is possible for TM modes, but this will not be done here.
- Note that the cutoff frequencies ω_{mn} for the TM modes are the same as those for TE modes.

Phase and Group Velocities

Phase Velocity and Group Velocity

- Phase velocity: simply the speed at which the crest of the wavefront travels in a given direction.
- For a plane wave $Ae^{i(\vec{k}.\vec{x}-\omega t)}$ the phase velocity along the direction $\hat{r}$ is

$$v_{ph} = \left.\frac{dr}{dt}\right|_{\text{constant phase}} = \frac{\omega}{|\vec{k}.\hat{r}|}$$

- If $\vec{k}$ is not along $\hat{r}$, typically $v_{ph} > c$ This does not mean that any signal is travelling faster than light, though.

- Group velocity measures the speed at which a signal is transported. The signal is embedded in the distribution of frequencies, and group velocity measures how fast the peak of this distribution shifts.

Group Velocity

- The Fourier transform of a wave gives the frequencies the wave consists of. Consider the situation where the spread in frequencies is small, which is the only one where we can define a group velocity easily. Let the frequencies be confined to the range $\omega = \omega_0 \pm \Delta\omega$. The corresponding wave vectors are confined to $\vec{\mathbf{k}} = \vec{\mathbf{k}}_0 \pm \Delta\vec{\mathbf{k}}$
- The wave is

$$\psi(\vec{x},t) = \int a(\vec{\mathbf{k}}) e^{i(\vec{\mathbf{k}},\vec{\mathbf{x}}-\omega t)} d^3k,$$

which may be written as

$$\psi(\vec{\mathbf{x}},t) = A(\vec{\mathbf{x}},t) e^{i(\vec{\mathbf{k}}_0 \vec{\mathbf{x}}-\omega_0 t)},$$

where

$$A(\vec{\mathbf{x}},t) = \int a(\vec{\mathbf{k}}) e^{i(\Delta\vec{\mathbf{k}},\vec{\mathbf{x}}-\Delta\omega t)} d^3k$$

- The frequency distribution shifts as a wavepacket, the velocity of the peak of the distribution is the approximate velocity of the wavepacket.
- Let us consider a one-dimensional case of a wave travelling along z-axis. At the peak,

$$0 = \frac{dA}{dt} = \frac{\partial A}{\partial t} + \frac{\partial A}{\partial z}\frac{dz}{dt}$$

- The group velocity is then

$$v_g = \frac{dz}{dt} = \frac{\partial A/\partial t}{\partial A/\partial z} = \frac{\Delta\omega}{\Delta k} = \left.\frac{d\omega}{dk}\right|_{\omega_0}$$

Velocities Along z Axis for the Waveguide

- $\omega = \sqrt{\omega_{mn}^2 + k_z^2}$
- Phase velocity $v_{ph} = \dfrac{\omega}{k_z} = \dfrac{c}{\sqrt{1-(\omega_{mn}/\omega)^2}}$
- Group velocity $v_g = \dfrac{d\omega}{dk_z} = c\sqrt{1-(\omega_{mn}/\omega)^2}$
- Waveguide transports different frequencies at different speeds: dispersion

Circular Cylindrical Waveguides

For TM Mode

$$E_z = AJ_m(k_\ell r)e^{im\phi}e^{i(k_z z-\omega t)}$$

If the cylinder has radius r_0, then the boundary condition is $J_m(k_\ell r_0)=0$ gives $k_\ell(m)$

For TM Mode

$$B_z = AJ_m(k_\ell r)e^{im\phi}e^{i(k_z z-\omega t)}$$

If the cylinder has radius r_0, then the boundary condition is $J_m(k_\ell r_0)=0$ gives $k_\ell(m)$

- $k_z^2 = (\omega/c)^2 - k_\ell^2 \Rightarrow$ cutoff frequency $\omega_{m,\ell} = ck_\ell(m)$
- TM and TE modes have different cutoff frequencies, unlike rectangular waveguides !

Power Transmitted by a Waveguide

- Consider TE mode. i.e. $E_z = 0$.
- The equations for $\vec{\mathbf{E}}_\perp = (E_x, E_y)$ and $\vec{\mathbf{B}}_\perp = (B_x, B_y)$ become

$$\left.\begin{aligned} B_x &= \frac{ik_z}{k_\perp^2}\frac{\partial B_z}{\partial x} \\ B_y &= \frac{ik_z}{k_\perp^2}\frac{\partial B_z}{\partial y} \end{aligned}\right\} \Rightarrow \vec{\mathbf{B}}_\perp = \frac{ik_z}{k_\perp^2}\nabla_\perp B_z$$

$$\left.\begin{aligned} E_x &= \frac{ck}{k_\perp^2}\frac{\partial B_z}{\partial y} \\ E_y &= -\frac{ck}{k_\perp^2}\frac{\partial B_z}{\partial x} \end{aligned}\right\} \Rightarrow \vec{\mathbf{E}}_\perp = \frac{ick}{k_z}\vec{\mathbf{B}}_\perp \times \vec{\mathbf{z}}$$

- The magnitude of Poynting vector (power transmitted per unit area) is then

$$\left|\vec{\mathbf{N}}\right| = \frac{\left|\vec{\mathbf{E}}^*_{\perp 0} \times \vec{\mathbf{H}}_{\perp 0}\right|^2}{2} = \frac{\left|\vec{\mathbf{E}}^*_{\perp 0}\right|^2}{2}\vec{\mathbf{k}}_z ck_{\mu 0} = \frac{1}{2}\sqrt{\frac{\varepsilon_0}{\mu_0}}\frac{k_z}{k}\left|\vec{\mathbf{E}}_0\right|^2$$

- Comparing with $\left|\vec{\mathbf{N}}\right| = (1/2)\sigma\left|E_0\right|^2$ this enables us to define the conductance of the waveguide as $\sigma = \sqrt{\varepsilon_0/\mu_0}\,(k_z/k)$. This may be compared with the conductance of free space, $\sqrt{\varepsilon_0/\mu_0}$.

Coaxial Cable

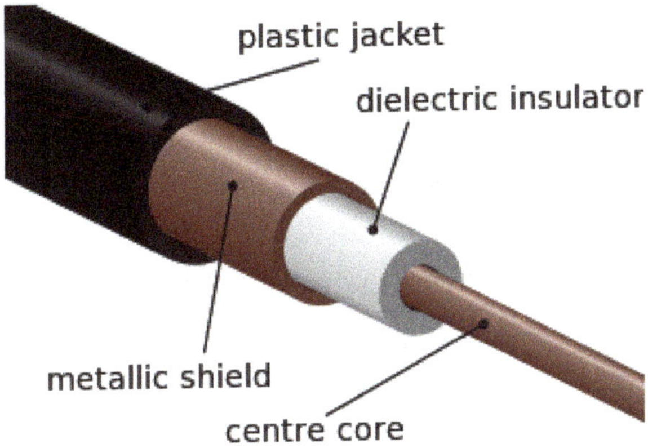
plastic jacket
dielectric insulator
metallic shield
centre core

Propagation Through a Coaxial Cable

- TEM Mode is supported (now there are two disjoint boundaries, so the argument for hollow waveguides does not work.)
- TE and TM modes also propagate, but have a threshold frequency

The TEM Mode

- Electric and magnetic fields:

$$\vec{\mathbf{E}} = \frac{E_0 \hat{r}}{r} e^{i(k_z z - \omega t)}, \vec{\mathbf{B}} = \frac{E_0 \hat{\phi}}{cr} e^{i(k_z z - \omega t)}$$

- Group velocity $V_g = c$

Cavities

LHC Accelerator: Cavity Principle

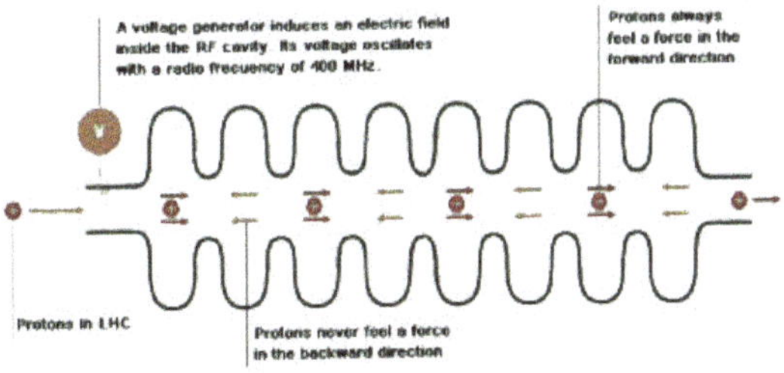
A voltage generator induces an electric field inside the RF cavity. Its voltage oscillates with a radio frecuency of 400 MHz.
Protons always feel a force in the forward direction
Protons in LHC
Protons never feel a force in the backward direction

Rectangular Cavity

- Conducting walls at x = 0, a; at y = 0, b and at z = 0, c.
- Potential inside the cavity:

$$\Phi_{mnp} = \sin(k_x x)\sin(k_y y)\sin(k_z z)e^{-i\omega t}$$

where $k_x = (m\pi/a)$, $k_y = (n\pi/b)$, $k_z = (p\pi/c)$

- This can be used to obtain $\vec{\mathbf{E}}$ and $\vec{\mathbf{B}}$ inside the cavity.
- A rectangular cavity supports discrete modes.

LHC Accelerator: Bunching Cavities

References

- Quantum Physics of Atoms, Molecules, Solids, Nuclei, and Particles (2nd Edition), R. Eisberg, R. Resnick, John Wiley & Sons, 1985, ISBN 978-0-471-87373-0
- Hiebert, W; Ballentine, G; Freeman, M (2002). "Comparison of experimental and numerical micromagnetic dynamics in coherent precessional switching and modal oscillations". Physical Review B. 65 (14). p. 140404. doi:10.1103/PhysRevB.65.140404
- Ramo, Simon; Whinnery, John R.; Van Duzer, Theodore (1994). Fields and Waves in Communication Electronics. New York: Joh Wiley and Sons. pp. 321–324. ISBN 0-471-58551-3
- Kuhn, U.; Lüty, F. (1965). "Paraelectric heating and cooling with OH—dipoles in alkali halides". Solid State Communications. 3 (2): 31. Bibcode:1965SSCom...3...31K. doi:10.1016/0038-1098(65)90060-8
- Oliner, Arthur A, "The evolution of electromagnetic waveguides: from hollow metallic guides to microwave integrated circuits", chapter 16 in, Sarkar et al., History of Wireless, Wiley, 2006 ISBN 0471783013
- Lyon, David (2013). "Gap size dependence of the dielectric strength in nano vacuum gaps". IEEE Transactions on Dielectrics and Electrical Insulation. 20 (4). doi:10.1109/TDEI.2013.6571470

3

Relativity and Electrodynamics

Lorentz force can be understood as the force on an electric charge q moving with velocity v in a magnetic field B and electric field E. Lorentz force, when taken with Maxwell's equation, establishes the underlying principles of electric, optical and radio technologies. This chapter discusses in detail the theories and methodologies related to relativity and electrodynamics.

Lorentz Force

In physics (particularly in electromagnetism) the Lorentz force is the combination of electric and magnetic force on a point charge due to electromagnetic fields. A particle of charge *q* moving with velocity v in the presence of an electric field E and a magnetic field B experiences a force

$$\mathbf{F} = q\mathbf{E} + q\mathbf{v} \times \mathbf{B}$$

(in IS units). Variations on this basic formula describe the magnetic force on a current-carrying wire (sometimes called Laplace force), the electromotive force in a wire loop moving through a magnetic field (an aspect of Faraday's law of induction), and the force on a charged particle which might be travelling near the speed of light (relativistic form of the Lorentz force).

The first derivation of the Lorentz force is commonly attributed to Oliver Heaviside in 1889, although other historians suggest an earlier origin in an 1865 paper by James Clerk Maxwell. Hendrik Lorentz derived it a few years after Heaviside.

Equation

Charged Particle

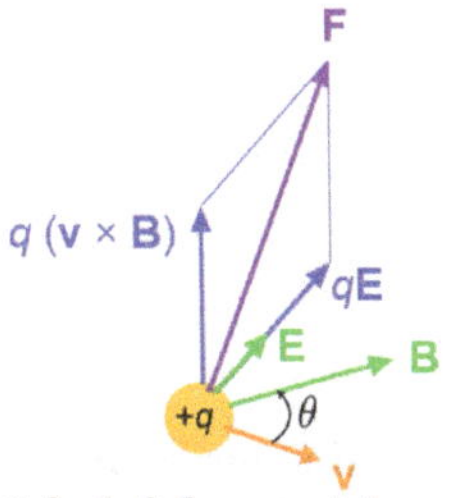

Lorentz force F on a charged particle (of charge *q*) in motion (instantaneous velocity v). The E field and B field vary in space and time.

The force F acting on a particle of electric charge *q* with instantaneous velocity v, due to an external electric field E and magnetic field B, is given by (in SI units):

$$\mathbf{F} = q(\mathbf{E} + \mathbf{v} \times \mathbf{B})$$

where × is the vector cross product. All boldface quantities are vectors. More explicitly stated:

$$\mathbf{F}(\mathbf{r}, \dot{\mathbf{r}}, t, q) = q[\mathbf{E}(\mathbf{r}, t) + \dot{\mathbf{r}} \times \mathbf{B}(\mathbf{r}, t)]$$

in which r is the position vector of the charged particle, *t* is time, and the overdot is a time derivative.

A positively charged particle will be accelerated in the *same* linear orientation as the E field, but will curve perpendicularly to both the instantaneous velocity vector v and the B field according to the right-hand rule (in detail, if the fingers of the right hand are extended to point in the direction of v and are then curled to point in the direction of B, then the extended thumb will point in the direction of F).

The term *q*E is called the electric force, while the term *q*v × B is called the magnetic force. According to some definitions, the term "Lorentz force" refers specifically to the formula for the magnetic force, with the *total* electromagnetic force (including the electric force) given some other (nonstandard) name. This section will *not* follow this nomenclature: In what follows, the term "Lorentz force" will refer only to the expression for the total force.

The magnetic force component of the Lorentz force manifests itself as the force that acts on a current-carrying wire in a magnetic field. In that context, it is also called the Laplace force.

Continuous Charge Distribution

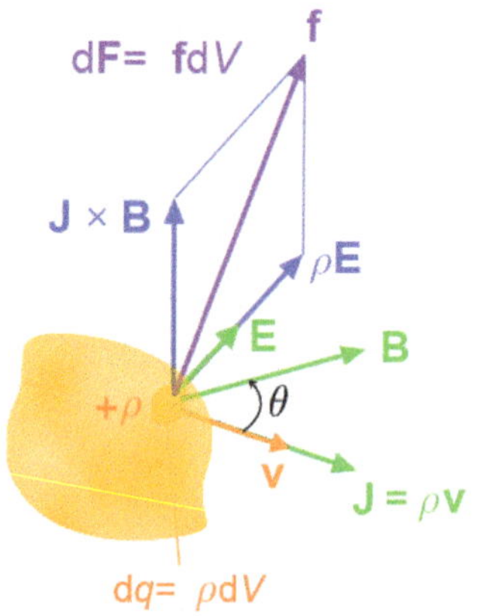

Lorentz force (per unit 3-volume) f on a continuous charge distribution (charge density ρ) in motion. The 3-current density J corresponds to the motion of the charge element *dq* in volume element *dV* and varies throughout the continuum.

For a continuous charge distribution in motion, the Lorentz force equation becomes:

$$d\mathbf{F} = dq\left(\mathbf{E} + \mathbf{v} \times \mathbf{B}\right)$$

where *dF* is the force on a small piece of the charge distribution with charge *dq*. If both sides of this equation are divided by the volume of this small piece of the charge distribution *dV*, the result is:

$$\mathbf{f} = \rho\left(\mathbf{E} + \mathbf{v} \times \mathbf{B}\right)$$

where f is the *force density* (force per unit volume) and ρ is the charge density (charge per unit volume). Next, the current density corresponding to the motion of the charge continuum is

$$\mathbf{J} = \rho\mathbf{v}$$

so the continuous analogue to the equation is

$$\mathbf{f} = \rho\mathbf{E} + \mathbf{J} \times \mathbf{B}$$

The total force is the volume integral over the charge distribution:

$$\mathbf{F} = \iiint (\rho\mathbf{E} + \mathbf{J} \times \mathbf{B})dV.$$

By eliminating ρ and J, using Maxwell's equations, and manipulating using the theorems of vector calculus, this form of the equation can be used to derive the Maxwell stress tensor σ, in turn this can be combined with the Poynting vector S to obtain the electromagnetic stress–energy tensor T used in general relativity.

In terms of σ and S, another way to write the Lorentz force (per unit volume) is

$$\mathbf{f} = \nabla \cdot \sigma - \frac{1}{c^2}\frac{\partial \mathbf{S}}{\partial t}$$

where *c* is the speed of light and ∇· denotes the divergence of a tensor field. Rather than the amount of charge and its velocity in electric and magnetic fields, this equation relates the energy flux (flow of *energy* per unit time per unit distance) in the fields to the force exerted on a charge distribution.

Equation in Cgs Units

The above-mentioned formulae use SI units which are the most common among experimentalists, technicians, and engineers. In cgs-Gaussian units, which are somewhat more common among theoretical physicists, one has instead

$$\mathbf{F} = q_{cgs}\left(\mathbf{E}_{cgs} + \frac{\mathbf{v}}{c}\times\mathbf{B}_{cgs}\right).$$

where c is the speed of light. Although this equation looks slightly different, it is completely equivalent, since one has the following relations:

$$q_{cgs} = \frac{q_{SI}}{\sqrt{4\pi\epsilon_0}}, \quad \mathbf{E}_{cgs} = \sqrt{4\pi\epsilon_0}\,\mathbf{E}_{SI}, \quad \mathbf{B}_{cgs} = \sqrt{4\pi/\mu_0}\,\mathbf{B}_{SI}, \quad c = \frac{1}{\sqrt{\epsilon_0\mu_0}}.$$

where ε_0 is the vacuum permittivity and μ_0 the vacuum permeability. In practice, the subscripts "cgs" and "SI" are always omitted, and the unit system has to be assessed from context.

History

Early attempts to quantitatively describe the electromagnetic force were made in the mid-18th century. It was proposed that the force on magnetic poles, by Johann Tobias Mayer and others in 1760,and electrically charged objects, by Henry Cavendish in 1762, obeyed an inverse-square law. However, in both cases the experimental proof was neither complete nor conclusive. It was not until 1784 when Charles-Augustin de Coulomb, using a torsion balance, was able to definitively show through experiment that this was true. Soon after the discovery in 1820 by H. C. Ørsted that a magnetic needle is acted on by a voltaic current, André-Marie Ampère that same year was able to devise through experimentation the formula for the angular dependence of the force between two current elements. In all these descriptions, the force was always given in terms of the properties of the objects involved and the distances between them rather than in terms of electric and magnetic fields.

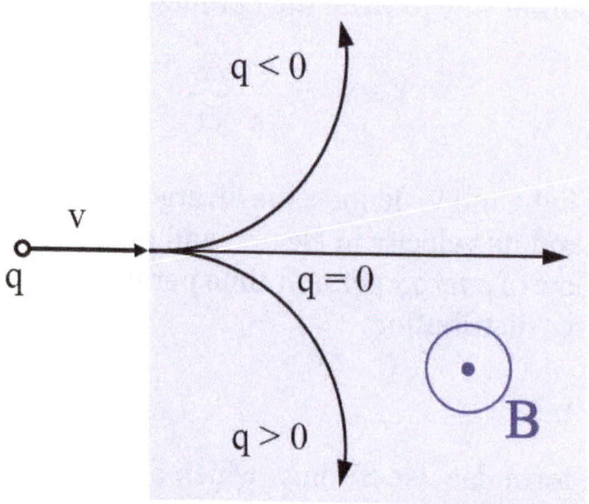

Trajectory of a particle with a positive or negative charge *q* under the influence of a magnetic field *B*, which is directed perpendicularly out of the screen.

Beam of electrons moving in a circle, due to the presence of a magnetic field. Purple light is emitted along the electron path, due to the electrons colliding with gas molecules in the bulb. A Teltron tube is used in this example.

The modern concept of electric and magnetic fields first arose in the theories of Michael Faraday, particularly his idea of lines of force, later to be given full mathematical description by Lord Kelvin and James Clerk Maxwell. From a modern perspective it is possible to identify in Maxwell's 1865 formulation of his field equations a form of the Lorentz force equation in relation to electric currents, however, in the time of Maxwell it was not evident how his equations related to the forces on moving charged objects. J. J. Thomson was the first to attempt to derive from Maxwell's field equations the electromagnetic forces on a moving charged object in terms of the object's properties and external fields. Interested in determining the electromagnetic behavior of the charged particles in cathode rays, Thomson published a paper in 1881 wherein he gave the force on the particles due to an external magnetic field as

$$\mathbf{F} = \frac{q}{2}\mathbf{v} \times \mathbf{B}.$$

Thomson derived the correct basic form of the formula, but, because of some miscalculations and an incomplete description of the displacement current, included an incorrect scale-factor of a half in front of the formula. Oliver Heaviside invented the modern vector notation and applied it to Maxwell's field equations; he also (in 1885 and 1889) had fixed the mistakes of Thomson's derivation and arrived at the correct form of the magnetic force on a moving charged object. Finally, in 1892, Hendrik Lorentz derived the modern form of the formula for the electromagnetic force which includes the contributions to the total force from both the electric and the magnetic fields. Lorentz began by abandoning the Maxwellian descriptions of the ether and conduction. Instead, Lorentz made a distinction between matter and the luminiferous aether and sought to apply the Maxwell equations at a microscopic scale. Using Heaviside's version of the Maxwell equations for a stationary ether and applying Lagrangian mechanics, Lorentz arrived at the correct and complete form of the force law that now bears his name.

Trajectories of Particles Due to the Lorentz Force

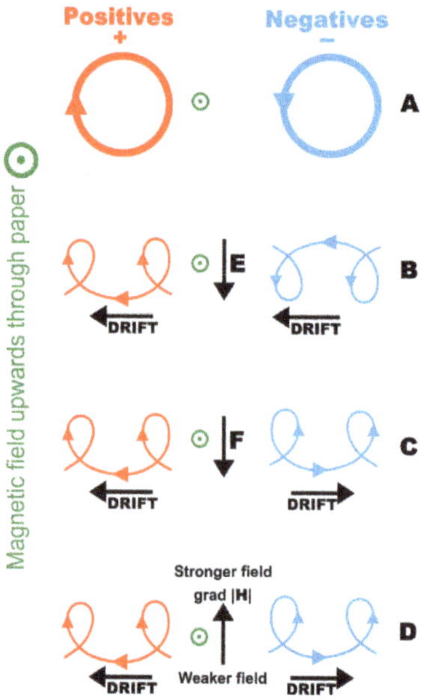

Charged particle drifts in a homogeneous magnetic field. (A) No disturbing force (B) With an electric field, E (C) With an independent force, F (e.g. gravity) (D) In an inhomogeneous magnetic field, grad H

In many cases of practical interest, the motion in a magnetic field of an electrically charged particle (such as an electron or ion in a plasma) can be treated as the superposition of a relatively fast circular motion around a point called the guiding center and a relatively slow drift of this point. The drift speeds may differ for various species depending on their charge states, masses, or temperatures, possibly resulting in electric currents or chemical separation.

Significance of the Lorentz Force

While the modern Maxwell's equations describe how electrically charged particles and currents or moving charged particles give rise to electric and magnetic fields, the Lorentz force law completes that picture by describing the force acting on a moving point charge q in the presence of electromagnetic fields. The Lorentz force law describes the effect of E and B upon a point charge, but such electromagnetic forces are not the entire picture. Charged particles are possibly coupled to other forces, notably gravity and nuclear forces. Thus, Maxwell's equations do not stand separate from other physical laws, but are coupled to them via the charge and current densities. The response of a point charge to the Lorentz law is one aspect; the generation of E and B by currents and charges is another.

In real materials the Lorentz force is inadequate to describe the collective behavior of charged particles, both in principle and as a matter of computation. The charged particles in a material medium not only respond to the E and B fields but also generate these fields. Complex transport equations must be solved to determine the time and spatial response of charges, for example, the Boltzmann equation or the Fokker–Planck equation or the Navier–Stokes equations. For example, magnetohydrodynamics, fluid dynamics, electrohydrodynamics, superconductivity, stellar evolution. An entire physical apparatus for dealing with these matters has developed.

Lorentz Force Law as the Definition of E and B

In many textbook treatments of classical electromagnetism, the Lorentz force Law is used as the *definition* of the electric and magnetic fields E and B. To be specific, the Lorentz force is understood to be the following empirical statement:

> *The electromagnetic force **F** on a test charge at a given point and time is a certain function of its charge* q *and velocity **v**, which can be parameterized by exactly two vectors **E** and **B**, in the functional form:*

$$\mathbf{F} = q(\mathbf{E} + \mathbf{v} \times \mathbf{B})$$

This is valid, even for particles approaching the speed of light (that is, magnitude of $v = |v| = c$). So the two vector fields E and B are thereby defined throughout space and time, and these are called the "electric field" and "magnetic field". The fields are defined everywhere in space and time with respect to what force a test charge would receive regardless of whether a charge is present to experience the force.

As a definition of E and B, the Lorentz force is only a definition in principle because a real particle (as opposed to the hypothetical "test charge" of infinitesimally-small mass and charge) would generate its own finite E and B fields, which would alter the electromagnetic force that it experiences. In addition, if the charge experiences acceleration, as if forced into a curved trajectory by some external agency, it emits radiation that causes braking of its motion. These effects occur through both a direct effect (called the radiation reaction force) and indirectly (by affecting the motion of nearby charges and currents). Moreover, net force must include gravity, electroweak, and any other forces aside from electromagnetic force.

Force on a Current-carrying Wire

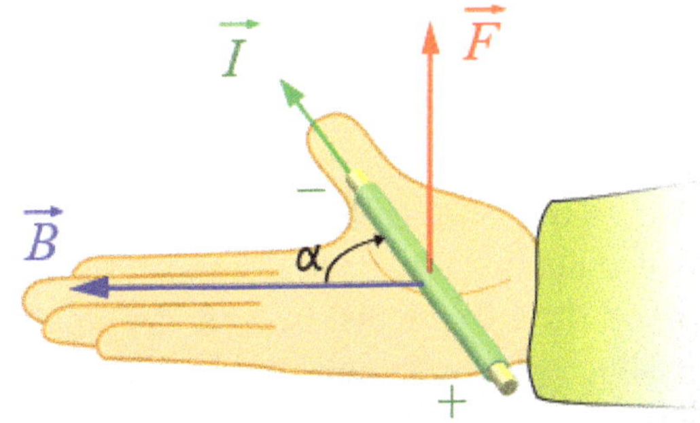

Right-hand rule for a current-carrying wire in a magnetic field *B*

When a wire carrying an electric current is placed in a magnetic field, each of the moving charges, which comprise the current, experiences the Lorentz force, and together they can create a macroscopic force on the wire (sometimes called the Laplace force). By combining the Lorentz force law above with the definition of electric current, the following equation results, in the case of a straight, stationary wire:

$$\mathbf{F} = I\ell \times \mathbf{B}$$

where ℓ is a vector whose magnitude is the length of wire, and whose direction is along the wire, aligned with the direction of conventional current flow *I*.

If the wire is not straight but curved, the force on it can be computed by applying this formula to each infinitesimal segment of wire $d\ell$, then adding up all these forces by integration. Formally, the net force on a stationary, rigid wire carrying a steady current *I* is

$$\mathbf{F} = I\int d\ell \times \mathbf{B}$$

This is the net force. In addition, there will usually be torque, plus other effects if the wire is not perfectly rigid.

One application of this is Ampère's force law, which describes how two current-carrying wires can attract or repel each other, since each experiences a Lorentz force from the other's magnetic field.

EMF

The magnetic force ($qv \times B$) component of the Lorentz force is responsible for *motional* electromotive force (or *motional EMF*), the phenomenon underlying many electrical generators. When a conductor is moved through a magnetic field, the magnetic field exerts opposite forces on electrons and nuclei in the wire, and this creates the EMF. The term "motional EMF" is applied to this phenomenon, since the EMF is due to the *motion* of the wire.

In other electrical generators, the magnets move, while the conductors do not. In this case, the EMF is due to the electric force (qE) term in the Lorentz Force equation. The electric field in question is created by the changing magnetic field, resulting in an *induced* EMF, as described by the Maxwell–Faraday equation (one of the four modern Maxwell's equations).

Both of these EMFs, despite their apparently distinct origins, are described by the same equation, namely, the EMF is the rate of change of magnetic flux through the wire. Einstein's special theory of relativity was partially motivated by the desire to better understand this link between the two effects. In fact, the electric and magnetic fields are different facets of the same electromagnetic field, and in moving from one inertial frame to another, the solenoidal vector field portion of the *E*-field can change in whole or in part to a *B*-field or *vice versa*.

Lorentz Force and Faraday's Law of Induction

Lorentz force -image on a wall in Leiden

Given a loop of wire in a magnetic field, Faraday's law of induction states the induced electromotive force (EMF) in the wire is:

$$\mathcal{E} = -\frac{d\Phi_B}{dt}$$

where

$$\Phi_B = \iint_{\Sigma(t)} d\mathbf{A} \cdot \mathbf{B}(\mathbf{r}, t)$$

is the magnetic flux through the loop, B is the magnetic field, $\Sigma(t)$ is a surface bounded by the closed contour $\partial\Sigma(t)$, at all at time t, dA is an infinitesimal vector area element of $\Sigma(t)$ (magnitude is the area of an infinitesimal patch of surface, direction is orthogonal to that surface patch).

The *sign* of the EMF is determined by Lenz's law. Note that this is valid for not only a *stationary* wire – but also for a *moving* wire.

From Faraday's law of induction (that is valid for a moving wire, for instance in a motor) and the Maxwell Equations, the Lorentz Force can be deduced. The reverse is also true, the Lorentz force and the Maxwell Equations can be used to derive the Faraday Law.

Let $\Sigma(t)$ be the moving wire, moving together without rotation and with constant velocity v and $\Sigma(t)$ be the internal surface of the wire. The EMF around the closed path $\partial\Sigma(t)$ is given by:

$$\mathcal{E} = \oint_{\partial\Sigma(t)} d\ell \cdot \mathbf{F} / q$$

where

$$\mathbf{E} = \mathbf{F} / q$$

is the electric field and $d\ell$ is an infinitesimal vector element of the contour $\partial\Sigma(t)$.

NB: Both $d\ell$ and dA have a sign ambiguity; to get the correct sign, the right-hand rule is used, as explained in the article Kelvin–Stokes theorem.

The above result can be compared with the version of Faraday's law of induction that appears in the modern Maxwell's equations, called here the *Maxwell–Faraday equation*:

$$\nabla \times \mathbf{E} = -\frac{\partial \mathbf{B}}{\partial t}.$$

The Maxwell–Faraday equation also can be written in an *integral form* using the Kelvin–Stokes theorem.

So we have, the Maxwell Faraday equation:

$$\oint_{\partial\Sigma(t)} d\ell \cdot \mathbf{E}(\mathbf{r}, t) = -\iint_{\Sigma(t)} d\mathbf{A} \cdot \frac{d\mathbf{B}(\mathbf{r}, t)}{dt}$$

and the Faraday Law,

$$\oint_{\partial\Sigma(t)} d\ell \cdot \mathbf{F} / q(\mathbf{r}, t) = -\frac{d}{dt} \iint_{\Sigma(t)} d\mathbf{A} \cdot \mathbf{B}(\mathbf{r}, t).$$

The two are equivalent if the wire is not moving. Using the Leibniz integral rule and that *div* B = 0, results in,

$$\oint_{\partial\Sigma(t)} d\ell \cdot \mathbf{F} / q(\mathbf{r}, t) = -\iint_{\Sigma(t)} d\mathbf{A} \cdot \frac{\partial}{\partial t} \mathbf{B}(\mathbf{r}, t) + \oint_{\partial\Sigma(t)} \mathbf{v} \times \mathbf{B} d\ell$$

and using the Maxwell Faraday equation,

$$\oint_{\partial\Sigma(t)} d\ell \cdot \mathbf{F} / q(\mathbf{r}, t) = \oint_{\partial\Sigma(t)} d\ell \cdot \mathbf{E}(\mathbf{r}, t) + \oint_{\partial\Sigma(t)} \mathbf{v} \times \mathbf{B}(\mathbf{r}, t) d\ell$$

since this is valid for any wire position it implies that,

$$\mathbf{F} = q\mathbf{E}(\mathbf{r}, t) + q\mathbf{v} \times \mathbf{B}(\mathbf{r}, t).$$

Faraday's law of induction holds whether the loop of wire is rigid and stationary, or in motion or in process of deformation, and it holds whether the magnetic field is constant in time or changing. However, there are cases where Faraday's law is either inadequate or difficult to use, and application of the underlying Lorentz force law is necessary.

If the magnetic field is fixed in time and the conducting loop moves through the field, the magnetic flux Φ_B linking the loop can change in several ways. For example, if the B-field varies with position, and the loop moves to a location with different B-field, Φ_B will change. Alternatively, if the loop changes orientation with respect to the B-field, the B · dA differential element will change because of the different angle between B and d , also changing Φ_B. As a third example, if a portion of the circuit is swept through a uniform, time-independent B-field, and another portion of the circuit is held stationary, the flux linking the entire closed circuit can change due to the shift in relative position of the circuit's component parts with time (surface $\partial\Sigma(t)$ time-dependent). In all three cases, Faraday's law of induction then predicts the EMF generated by the change in Φ_B.

Note that the Maxwell Faraday's equation implies that the Electric Field E is non conservative when the Magnetic Field B varies in time, and is not expressible as the gradient of a scalar field, and not subject to the gradient theorem since its rotational is not zero.

Lorentz Force in Terms of Potentials

The E and B fields can be replaced by the magnetic vector potential A and (scalar) electrostatic potential ϕ by

$$\mathbf{E} = -\nabla\phi - \frac{\partial \mathbf{A}}{\partial t}$$

$$\mathbf{B} = \nabla \times \mathbf{A}$$

where ∇ is the gradient, $\nabla\cdot$ is the divergence, $\nabla\times$ is the curl.

The force becomes

$$\mathbf{F} = q\left[-\nabla\phi - \frac{\partial \mathbf{A}}{\partial t} + \mathbf{v} \times (\nabla \times \mathbf{A})\right]$$

and using an identity for the triple product simplifies to

$$\mathbf{F} = q\left[-\nabla\phi - \frac{\partial \mathbf{A}}{\partial t} + \nabla(\mathbf{v}\cdot\mathbf{A}) - (\mathbf{v}\cdot\nabla)\mathbf{A}\right]$$

using the chain rule, the total derivative of A is:

$$\frac{d\mathbf{A}}{dt} = \frac{\partial \mathbf{A}}{\partial t} + (\mathbf{v}\cdot\nabla)\mathbf{A}$$

so the above expression can be rewritten as;

$$\mathbf{F} = q\left[-\nabla(\phi - \mathbf{v}\cdot\mathbf{A}) - \frac{d\mathbf{A}}{dt}\right]$$

which can take the convenient Euler–Lagrange form

$$\mathbf{F} = q\left[-\nabla_x(\phi - \dot{x}\cdot\mathbf{A}) + \frac{d}{dt}\nabla_{\dot{x}}(\phi - \dot{x}\cdot\mathbf{A})\right]$$

Lorentz Force and Analytical Mechanics

The Lagrangian for a charged particle of mass *m* and charge *q* in an electromagnetic field equivalently describes the dynamics of the particle in terms of its *energy*, rather than the force exerted on it. The classical expression is given by:

$$L = \frac{m}{2}\dot{\mathbf{r}}\cdot\dot{\mathbf{r}} + q\mathbf{A}\cdot\dot{\mathbf{r}} - q\phi$$

where A and ϕ are the potential fields as above. Using Lagrange's equations, the equation for the Lorentz force can be obtained.

Derivation of Lorentz Force from Classical Lagrangian (SI units)

For an A field, a particle moving with velocity v = ṙ has potential momentum $q\mathbf{A}(\mathbf{r},t)$, so its potential energy is $q\mathbf{A}(\mathbf{r},t)\cdot\dot{\mathbf{r}}$. For a ϕ field, the particle's potential energy is $q\phi(\mathbf{r},t)$.

The total potential energy is then:

$$V = q\phi - q\mathbf{A}\cdot\dot{\mathbf{r}}$$

and the kinetic energy is:

$$T = \frac{m}{2}\dot{\mathbf{r}}\cdot\dot{\mathbf{r}}$$

hence the Lagrangian:

$$L = T - V = \frac{m}{2}\dot{\mathbf{r}}\cdot\dot{\mathbf{r}} + q\mathbf{A}\cdot\dot{\mathbf{r}} - q\phi$$

$$L = \frac{m}{2}(\dot{x}^2 + \dot{y}^2 + \dot{z}^2) + q(\dot{x}A_x + \dot{y}A_y + \dot{z}A_z) - q\phi$$

Lagrange's equations are

$$\frac{d}{dt}\frac{\partial L}{\partial \dot{x}} = \frac{\partial L}{\partial x}$$

(same for *y* and *z*). So calculating the partial derivatives:

$$\frac{d}{dt}\frac{\partial L}{\partial \dot{x}} = m\ddot{x} + q\frac{dA_x}{dt}$$

$$= m\ddot{x} + \frac{q}{dt}\left(\frac{\partial A_x}{\partial t}dt + \frac{\partial A_x}{\partial x}dx + \frac{\partial A_x}{\partial y}dy + \frac{\partial A_x}{\partial z}dz\right)$$

$$= m\ddot{x} + q\left(\frac{\partial A_x}{\partial t} + \frac{\partial A_x}{\partial x}\dot{x} + \frac{\partial A_x}{\partial y}\dot{y} + \frac{\partial A_x}{\partial z}\dot{z}\right)$$

$$\frac{\partial L}{\partial x} = -q\frac{\partial \phi}{\partial x} + q\left(\frac{\partial A_x}{\partial x}\dot{x} + \frac{\partial A_y}{\partial x}\dot{y} + \frac{\partial A_z}{\partial x}\dot{z}\right)$$

equating and simplifying:

$$m\ddot{x} + q\left(\frac{\partial A_x}{\partial t} + \frac{\partial A_x}{\partial x}\dot{x} + \frac{\partial A_x}{\partial y}\dot{y} + \frac{\partial A_x}{\partial z}\dot{z}\right) = -q\frac{\partial \phi}{\partial x} + q\left(\frac{\partial A_x}{\partial x}\dot{x} + \frac{\partial A_y}{\partial x}\dot{y} + \frac{\partial A_z}{\partial x}\dot{z}\right)$$

$$F_x = -q\left(\frac{\partial \phi}{\partial x} + \frac{\partial A_x}{\partial t}\right) + q\left[\dot{y}\left(\frac{\partial A_y}{\partial x} - \frac{\partial A_x}{\partial y}\right) + \dot{z}\left(\frac{\partial A_z}{\partial x} - \frac{\partial A_x}{\partial z}\right)\right]$$

$$= qE_x + q[\dot{y}(\nabla \times \mathbf{A})_z - \dot{z}(\nabla \times \mathbf{A})_y]$$

$$= qE_x + q[\dot{\mathbf{r}} \times (\nabla \times \mathbf{A})]_x$$

$$= qE_x + q(\dot{\mathbf{r}} \times \mathbf{B})_x$$

and similarly for the *y* and *z* directions. Hence the force equation is:

$$\mathbf{F} = q(\mathbf{E} + \dot{\mathbf{r}} \times \mathbf{B})$$

The potential energy depends on the velocity of the particle, so the force is velocity dependent, so it is not conservative.

The relativistic Lagrangian is

$$L = -mc^2\sqrt{1 - \left(\frac{\dot{r}}{c}\right)^2} + e\mathbf{A}(\mathbf{r})\cdot\dot{r} - e\phi(\mathbf{r})$$

The action is the relativistic arclength of the path of the particle in space time, minus the potential energy contribution, plus an extra contribution which quantum mechanically is an extra phase a charged particle gets when it is moving along a vector potential.

Derivation of Lorentz Force from Relativistic Lagrangian (SI units)

The equations of motion derived by extremizing the action:

$$\frac{d\mathbf{P}}{dt} = \frac{\partial L}{\partial \mathbf{r}} = e\frac{\partial \mathbf{A}}{\partial \mathbf{r}} \cdot \dot{\mathbf{r}} - e\frac{\partial \phi}{\partial \mathbf{r}}$$

$$\mathbf{P} - e\mathbf{A} = \frac{m\dot{\mathbf{r}}}{\sqrt{1-\left(\frac{\dot{\mathbf{r}}}{c}\right)^2}}$$

are the same as Hamilton's equations of motion:

$$\frac{d\mathbf{r}}{dt} = \frac{\partial}{\partial \mathbf{p}}\left(\sqrt{(\mathbf{P}-e\mathbf{A})^2 + (mc^2)^2} + e\phi\right)$$

$$\frac{d\mathbf{p}}{dt} = -\frac{\partial}{\partial \mathbf{r}}\left(\sqrt{(\mathbf{P}-e\mathbf{A})^2 + (mc^2)^2} + e\phi\right)$$

both are equivalent to the noncanonical form:

$$\frac{d}{dt}\left(\frac{m\dot{\mathbf{r}}}{\sqrt{1-\left(\frac{\dot{\mathbf{r}}}{c}\right)^2}}\right) = e\left(\mathbf{E} + \dot{\mathbf{r}} \times \mathbf{B}\right).$$

This formula is the Lorentz force, representing the rate at which the EM field adds relativistic momentum to the particle.

Relativistic Form of the Lorentz Force

Field Tensor

Using the metric signature $(1, -1, -1, -1)$, the Lorentz force for a charge q can be written in covariant form:

$$\frac{dp^\alpha}{d\tau} = qF^{\alpha\beta}U_\beta$$

where p^α is the four-momentum, defined as

$$p^{\alpha} = (p_0, p_1, p_2, p_3) = (\gamma mc, p_x, p_y, p_z),$$

τ the proper time of the particle, $F^{\alpha\beta}$ the contravariant electromagnetic tensor

$$F^{\alpha\beta} = \begin{pmatrix} 0 & -E_x/c & -E_y/c & -E_z/c \\ E_x/c & 0 & -B_z & B_y \\ E_y/c & B_z & 0 & -B_x \\ E_z/c & -B_y & B_x & 0 \end{pmatrix}$$

and U is the covariant 4-velocity of the particle, defined as:

$$U_{\beta} = (U_0, U_1, U_2, U_3) = \gamma(c, -v_x, -v_y, -v_z),$$

in which

$$\gamma(v) = \frac{1}{\sqrt{1 - \frac{v^2}{c^2}}} = \frac{1}{\sqrt{1 - \frac{v_x^2 + v_y^2 + v_z^2}{c^2}}}$$

is the Lorentz factor.

The fields are transformed to a frame moving with constant relative velocity by:

$$F'^{\mu\nu} = \Lambda^{\mu}{}_{\alpha} \Lambda^{\nu}{}_{\beta} F^{\alpha\beta},$$

where $\Lambda^{\mu}{}_{\alpha}$ is the Lorentz transformation tensor.

Translation to Vector Notation

The $\alpha = 1$ component (x-component) of the force is

$$\frac{dp^1}{d\tau} = qU_{\beta}F^{1\beta} = q\left(U_0F^{10} + U_1F^{11} + U_2F^{12} + U_3F^{13}\right).$$

Substituting the components of the covariant electromagnetic tensor F yields

$$\frac{dp^1}{d\tau} = q\left[U_0\left(\frac{E_x}{c}\right) + U_2(-B_z) + U_3(B_y)\right].$$

Using the components of covariant four-velocity yields

$$\frac{dp^1}{d\tau} = q\gamma\left[c\left(\frac{E_x}{c}\right) + (-v_y)(-B_z) + (-v_z)(B_y)\right]$$
$$= q\gamma\left(E_x + v_y B_z - v_z B_y\right)$$
$$= q\gamma\left[E_x + (\mathbf{v}\times\mathbf{B})_x\right].$$

The calculation for $\alpha = 2, 3$ (force components in the y and z directions) yields similar results, so collecting the 3 equations into one:

$$\frac{d\mathbf{p}}{d\tau} = q\gamma(\mathbf{E} + \mathbf{v}\times\mathbf{B}),$$

and since differentials in coordinate time dt and proper time $d\tau$ are related by the Lorentz factor,

$$dt = \gamma(v)d\tau,$$

so we arrive at

$$\frac{d\mathbf{p}}{dt} = q(\mathbf{E} + \mathbf{v}\times\mathbf{B}).$$

This is precisely the Lorentz force law, however, it is important to note that p is the relativistic expression,

$$\mathbf{p} = \gamma(v)m_0\mathbf{v}.$$

Lorentz Force in Spacetime Algebra (STA)

The electric and magnetic fields are dependent on the velocity of an observer, so the relativistic form of the Lorentz force law can best be exhibited starting from a coordinate-independent expression for the electromagnetic and magnetic fields $\mathcal{F}$, and an arbitrary time-direction, γ_0. This can be settled through Space-Time Algebra (or the geometric algebra of space-time), a type of Clifford's Algebra defined on a pseudo-euclidian space, as

$$\mathbf{E} = (\mathcal{F}\cdot\gamma_0)\gamma_0$$

and

$$i\mathbf{B} = (\mathcal{F}\wedge\gamma_0)\gamma_0$$

$\mathcal{F}$ is a space-time bivector (an oriented plane segment, just like a vector is an oriented line segment), which has six degrees of freedom corresponding to boosts (rotations in space-time planes) and rotations (rotations in space-space planes). The dot product with the vector γ_0 pulls a vector (in the space algebra) from the translational part, while the wedge-product creates a trivector (in the space algebra) who is dual to a vector which is the usual magnetic field vector. The relativistic velocity is given by the (time-like) changes in a time-position vector $v = \dot{x}$, where

$$v^2 = 1,$$

(which shows our choice for the metric) and the velocity is

$$\mathbf{v} = cv \wedge \gamma_0 / (v \cdot \gamma_0).$$

The proper (invariant is an inadequate term because no transformation has been defined) form of the Lorentz force law is simply

$$F = q\mathcal{F} \cdot v$$

Note that the order is important because between a bivector and a vector the dot product is anti-symmetric. Upon a space time split like one can obtain the velocity, and fields as above yielding the usual expression.

Faraday's Law and Lorentz Force

Faraday's Law of Induced EMF

Maxwell's Equations without External Sources

$$\nabla . \vec{\mathbf{E}} = 0 \qquad \nabla \times \vec{\mathbf{E}} = -\partial \vec{\mathbf{B}} / \partial t$$

$$\nabla . \vec{\mathbf{B}} = 0 \qquad \nabla \times \vec{\mathbf{B}} = \mu_0 \varepsilon_0 \partial \vec{\mathbf{E}} / \partial t$$

Faraday's Law

$$\nabla . \vec{\mathbf{E}} = -\partial \vec{\mathbf{B}} / \partial t \Rightarrow \oint \vec{\mathbf{E}} . d\ell = \partial \Phi / \partial t$$

- The total emf induced in a closed loop is equal to the rate of change of magnetic flux through the loop.
- Is this valid in all situations ?

Wire Loop Cutting Through Magnetic Field Lines

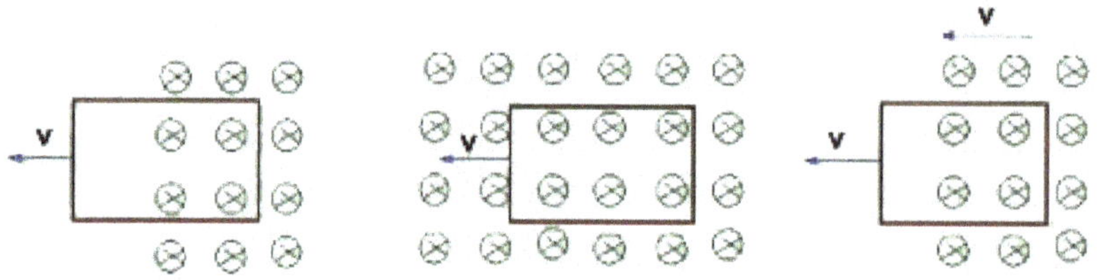

- In the first case, the magnetic flux through the loop clearly changes, so emf is induced in the wire loop, and current flows.
- In the second case, the wire cuts through the magnetic lines of force, but the total magnetic flux through the loop never changes. But current still flows ! (momentarily, when the loop starts moving, till the charge buildup cancels the EMF)
- In the third case, both the source of magnetic flux and the wire loop move together. The flux lines are not cut by the wire. But current still flows ! (and now, continues to flow)
- Maxwell's equations must be incomplete then...

Lorentz Force

Force on Charge q in the Presence of $\vec{\mathbf{E}}$ and $\vec{\mathbf{B}}$

$$\vec{\mathbf{F}} = q(\vec{\mathbf{E}} + \vec{\mathbf{v}} \times \vec{\mathbf{B}})$$

- This is an experimental result.
- It may be interpreted as an effective electric field:

$$\vec{\mathbf{E}}' = \vec{\mathbf{E}} + \vec{\mathbf{v}} \times \vec{\mathbf{B}}$$

- This relation is not contained in the Maxwell's equations (at this stage)
- The combination of two relations

$$\nabla \times \vec{\mathbf{E}} = -\partial \vec{\mathbf{B}} / \partial t, \text{ and } \vec{\mathbf{E}}' = \vec{\mathbf{E}} + \vec{\mathbf{v}} \times \vec{\mathbf{B}}$$

is sufficient to take care of all situations.

Faraday Disc: a Problem

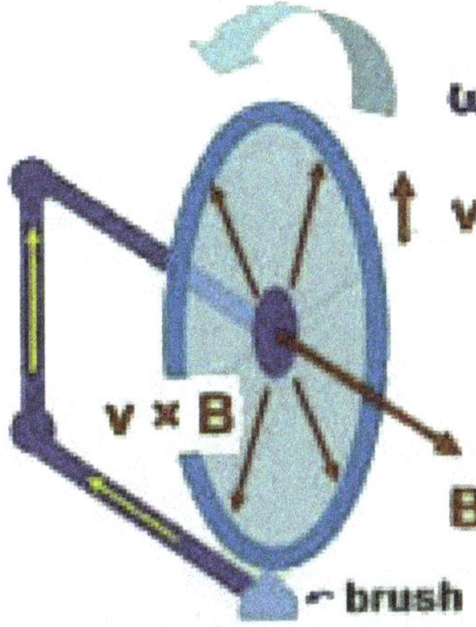

- Bar magnet along the axis of the conducting disc
- Conducting loop as shown in the figure

Motivations from Electrodynamics

- The Lorentz force law seems to be valid whether the conductor is moving in a magnetic field, or a magnet is moving near a conductor, only the relative speeds count.
- However with Galilean transformations $x' = x - vt$ and $t' = t$, the wave equation for a scalar potential $\Phi(x,t)$,

$$\left(\frac{\partial^2}{dx^2} - \frac{1}{c^2}\frac{\partial^2}{dt^2}\right)\Phi(x,t) = 0$$

  does not stay invariant. So an EM wave in one frame is not an EM wave in another frame?

- Indeed, the wave equation in vacuum for $\vec{\mathbf{E}}$ is

$$\nabla^2\vec{\mathbf{E}} + \mu_0\varepsilon_0(\partial^2\vec{\mathbf{E}}/\partial t^2) = 0$$

  leads to a wave travelling with speed $c = 1/\sqrt{\mu_0\varepsilon_0}$, which does not depend on the speed of the medium.

Motivations from Measurements of the Speed of Light

Michaelson-Morley Experiment (1887)

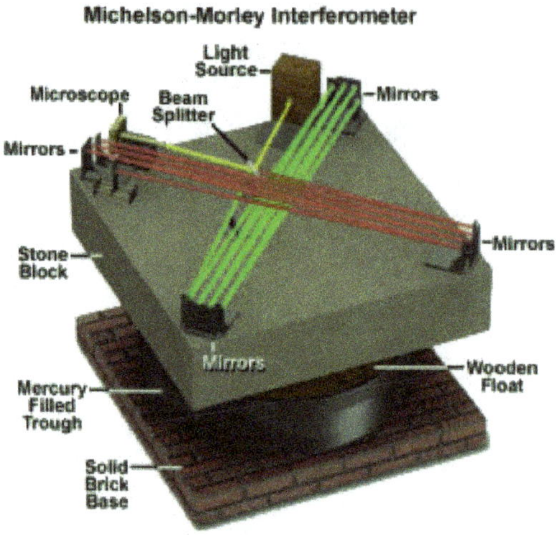

- Speed of light is independent of the speed of the medium through which light is travelling.
- Maxwell's equations also have been telling us the same thing !
- After the Michaelson-Morley experiment, FitzGerald wrote a terse paper in [The Ether and the earth's atmosphere, Science 13: 390 (1889)], postulating that lengths may be contracted along the direction of movement through ether.
- In 1892, Lorentz wrote a more quantitative paper, in which he calculated that contraction by the factor $1 - v^2/(2c^2)$, would explain the MM experiment. This is the "Lorentz FitzGerald contraction"
- Einstein's celebrated paper, used the consistency of Maxwell's equations as his main motivation to come up with the revolutionary concept that space and time are unified.

Lorentz Transformations of Observables

Length, Time, Velocity, Acceleration

Length Contraction

- The question: If the length of an object measured in frame S is L, what is the length measured in S'?

- Formal statement: An object is stationary in frame S. The coordinates of two ends of the object in this frame are x_1 and x_2, independent of t_1 or t_2. The measurement of length L corresponds to $x_2 - x_1 = L$. In frame S', the coordinates of these ends are (x_1', t_1') and (x_2', t_2'). The measurement of length in this frame corresponds to $L' = x_2' - x_1'$, when $t_1' = t_2'$.

- We have

$$x_1 = \gamma x_1' + \gamma\beta c t_1',\ x_2 = \gamma x_2' + \gamma\beta c t_2'$$

- This gives

$$x_2 - x_1 = \gamma(x_2' - x_1') + \gamma\beta c(t_2' - t_1')$$

- When $t_1' = t_2'$, one then gets

$$L' = L/\gamma$$

which is length contraction.

Time Dilation

- Formal statement: The time interval between two events at the same location in frame S (at $x_1 = x_2$) is $T = t_2 - t_1$, what is $t_2' - t_1'$1 in frame S' ?

- We have

$$ct_1' = \gamma c t_1 - \gamma\beta x_1,\ ct_2' = \gamma c t_2 - \gamma\beta x_2,$$

- Since $x_1 = x_2$, this gives

$$T' = (t_2' - t_1') = \gamma(t_2 - t_1) = \gamma T,$$

which is time dilation.

Velocity Measured from a Moving Frame

- Let the velocity of a particle in the frame S be u_x, u_y, u_z. What will be the velocity of this particle as measured in a frame S' moving with a speed v in the x-direction (in the S frame)?

- Formalizing: Given $dx/dt = u_x$, $dy/dt = u_y$, $dz/dt = u_z$,

determine $u_x' = dx'/dt'$, $u_y' = dy'/dt'$, $u_z' = dz'/dt'$.

- Using the Lorentz transformations

$$\begin{pmatrix} cdt' \\ dx' \\ dy' \\ dz' \end{pmatrix} = \begin{pmatrix} \gamma & -\gamma\beta & 0 & 0 \\ -\gamma\beta & \gamma & 0 & 0 \\ 0 & 0 & 1 & 0 \\ 0 & 0 & 0 & 1 \end{pmatrix} \begin{pmatrix} cdt \\ dx \\ dy \\ dz \end{pmatrix}$$

- One gets

$$u_x' = \frac{u_x - v}{1 - u_x v/c^2}, u_y' = \frac{u_y}{\gamma(1 - u_x v/c^2)}, u_z' = \frac{u_z}{\gamma(1 - u_x v/c^2)}$$

Velocity Addition

- The results about velocities measured in different frames lead directly to the "velocity addition" formula: If two particles A and B are moving towards each other with speeds u_A and u_B, then the speed of approach of A as measured by B is obtained simply by using $V = -u_B$ in the earlier results.

- This gives

$$U_A \oplus U_B = \frac{U_A + U_B}{1 + U_A U_B / c^2}$$

Note that this can never exceed c. Also, if either U_A or U_B equals c, then $U_A \oplus U_B = c$

Acceleration

- In frame S, a particle has instantaneous velocity (U_x, U_y, U_z) and acceleration (a_X, a_Y, a_Z). What is the velocity and acceleration in frame S' (moving with speed v along the x-direction) ?
- The velocity in the new frame is (U_x', U_y', U_z') as calculated earlier.
- The acceleration components are obtained simply through

$$a_x' = \frac{du_x'}{dt'}, a_y' = \frac{du_y'}{dt'}, a_z' = \frac{du_z'}{dt'}$$

where one has to use $du_x/dt = a_x$, $du_y/dt = a_y$, $du_z/dt = a_z$

Transformations of Electric and Magnetic Fields

Invariance of Maxwell's Equations

- We would like Maxwell's equations to be invariant under Lorentz transformations. That is, in vacuum, we would like to have

$$\nabla' \cdot \vec{\mathbf{E}}' = 0, \qquad \nabla' \times \vec{\mathbf{E}}' = -\partial \vec{\mathbf{B}}' / \partial t',$$

$$\nabla' \cdot \vec{\mathbf{B}}' = 0, \qquad \nabla' \times \vec{\mathbf{B}}' = \mu_0 \varepsilon_0 \partial \vec{\mathbf{E}}' / \partial t'$$

- This condition allows us to determine the transformation properties of $\vec{\mathbf{E}}$ and $\vec{\mathbf{B}}$.

- From the Lorentz transformations and the chain rule for derivatives,

$$\frac{\partial}{\partial(ct')} = \frac{\partial \mathrm{x}}{\partial(ct')}\frac{\partial}{\partial \mathrm{x}} + \frac{\partial(ct)}{\partial(ct')}\frac{\partial}{\partial(ct)} = \gamma\beta\frac{\partial}{\partial \mathrm{x}} + \gamma\frac{\partial}{\partial(ct)}$$

$$\frac{\partial}{\partial \mathrm{x}'} = \frac{\partial \mathrm{x}}{\partial \mathrm{x}'}\frac{\partial}{\partial \mathrm{x}} + \frac{\partial(ct)}{\partial \mathrm{x}'}\frac{\partial}{\partial(ct)} = \gamma\frac{\partial}{\partial \mathrm{x}} + \gamma\beta\frac{\partial}{\partial(ct)}$$

$$\frac{\partial}{\partial y'} = \frac{\partial}{\partial y}, \qquad \frac{\partial}{\partial z'} = \frac{\partial}{\partial z}$$

Transformations of $\vec{\mathbf{E}}$ and $\vec{\mathbf{B}}$ Fields

Problem

The transformations for components of ∇' and $\partial/\partial t'$ were obtained earlier in this section, in eq. 6. Assume the components of $\vec{\mathbf{E}}'$ and $\vec{\mathbf{B}}'$ to be some linear combinations of the components of $\vec{\mathbf{E}}$ and $\vec{\mathbf{B}}$, with coefficients that are functions of v. Show that

$$E'_{\mathrm{x}} = E_{\mathrm{x}}, \; E'_y = \gamma(E_y - vB_z), \; E'_z = \gamma(E_z + vB_y),$$

$$B'_{\mathrm{x}} = B_{\mathrm{x}}, \; B'_y = \gamma(B_y + \frac{v}{c^2}E_z), \; B'_z = \gamma(B_z - \frac{v}{c^2}E_y).$$

Lorentz Force from Relativistic Transformations

- We thus know how $\vec{\mathbf{E}}$ and $\vec{\mathbf{B}}$ fields should transform under a change of frame. The transformations can be written in the form:

$$\vec{\mathbf{E}}'_{\parallel} = \vec{\mathbf{E}}_{\parallel}, \quad \vec{\mathbf{E}}'_T = \gamma(\vec{\mathbf{E}}_T + \vec{\mathbf{v}} + \vec{\mathbf{B}}_T),$$

$$\vec{\mathbf{B}}'_{\parallel} = \vec{\mathbf{B}}_{\parallel}, \quad \vec{\mathbf{B}}'_T = \gamma(\vec{\mathbf{B}}_T + \frac{\vec{\mathbf{v}}}{c^2} + \vec{\mathbf{E}}_T),$$

where the subscript T denotes components transverse to the relative motion of frames, and || denotes components along the relative motion of frames.

- Note that in the limit of small velocities $(\gamma \approx 1)$, the above set of equations give

$$\vec{\mathbf{E}}' = \vec{\mathbf{E}}'_{\parallel} + \vec{\mathbf{E}}'_{T} \approx \vec{\mathbf{E}} + \vec{\mathbf{v}} \times \vec{\mathbf{B}}$$

the Lorentz force law. Thus, the principle of invariance of Maxwell's equations under Lorentz transformations has led us to the Lorentz force law directly.

More About the Lorentz Force

- Relativity tells us that Lorentz force law is not an arbitrary addition to Maxwell's equations, but just a consequence of the additional principle of invariance of Maxwell's equations under frame changes.
- It shows that the $\vec{\mathbf{E}}$ and $\vec{\mathbf{B}}$ fields are connected. A pure $\vec{\mathbf{B}}$ field in a frame give rise to an $\vec{\mathbf{E}}'$ field in another. On the other hand, magnetic fields $\vec{\mathbf{B}}'$ may be produced by purely electric fields $\vec{\mathbf{E}}$ in another inertial frame.
- The standard Lorentz force law is valid only at low velocities. At large velocities, the law "$\vec{\mathbf{E}}' = \vec{\mathbf{E}} + \vec{\mathbf{v}} \times \vec{\mathbf{B}}$" has to be modified to take care of the factors of γ in some components.

EM Wave: Aberration, Doppler Effect, Intensity

Invariance of the Wave Equation

- From the Lorentz transformations and the chain rule for derivatives,

$$\frac{\partial}{\partial(ct)} = \frac{\partial \mathrm{x}'}{\partial(ct)}\frac{\partial}{\partial \mathrm{x}'} + \frac{\partial(ct')}{\partial(ct)}\frac{\partial}{\partial(ct')} = -\gamma\beta\frac{\partial}{\partial \mathrm{x}'} + \gamma\frac{\partial}{\partial(ct')}$$

$$\frac{\partial}{\partial \mathrm{x}} = \frac{\partial \mathrm{x}'}{\partial \mathrm{x}}\frac{\partial}{\partial \mathrm{x}'} + \frac{\partial(ct')}{\partial \mathrm{x}}\frac{\partial}{\partial(ct')} = \gamma\frac{\partial}{\partial \mathrm{x}'} - \gamma\beta\frac{\partial}{\partial(ct')}$$

- This gives

$$\frac{\partial^2}{\partial(ct)^2} - \nabla^2 = \frac{\partial^2}{\partial(ct')^2} - \nabla'^2$$

so that the electromagnetic fields in free space, which are solutions of

$$\left(\frac{\partial^2}{\partial(ct)^2} - \nabla^2\right)\vec{\mathbf{V}}(\vec{\mathbf{x}}, t) = 0$$

are the solutions of

$$\left(\frac{\partial^2}{\partial(ct')^2}-\nabla'^2\right)\vec{\mathbf{V}}(\vec{\mathbf{x}}',t')=0$$

in any other inertial frame.

Invariance of the Plane Wave Solution

- Since the wave equation does not change, the plane wave solution in one frame stays a plane wave solution in another frame. That is, if a plane wave has the form

  $$\vec{\mathbf{V}}_0 \exp\left[i(\vec{\mathbf{k}}.\vec{\mathbf{x}}-\omega t)\right], \text{ with } \omega=c\left|\vec{\mathbf{k}}\right|$$

  in one frame S, it takes the form

  $$\vec{\mathbf{V}}'_0 \exp\left[i(\vec{\mathbf{k}}'.\vec{\mathbf{x}}'-\omega' t')\right], \text{ with } \omega'=c\left|\vec{\mathbf{k}}'\right|$$

  in the other frame S'.

- The relationship between the primed and unprimed values of k and ω can be obtained by equating the phases at $(\mathrm{x},\mathrm{y},\mathrm{z},\mathrm{t})$ and the corresponding $(\mathrm{x}',\mathrm{y}',\mathrm{z}',\mathrm{t}')$.

- If the frame S' is moxing with speed v along the x direction, then for a wave with $\vec{\mathbf{k}}=(k_x,k_y,k_z)$ and $\vec{\mathbf{k}}'=(k'_x,k'_y,k'_z)$, the phase-equality $\vec{\mathbf{k}}.\vec{\mathbf{x}}-\omega t=\vec{\mathbf{k}}'.\vec{\mathbf{x}}'-\omega' t'$ gives

  $$k_x\mathrm{x}+k_y y+k_z z-\left|\vec{\mathbf{k}}\right|ct=$$
  $$(\gamma k'_x+\gamma\beta\left|\vec{\mathbf{k}}'\right|)\mathrm{x}+k'_y y+k'_z z-(\gamma\beta k'_x+\gamma\left|\vec{\mathbf{k}}'\right|)ct.$$

Aberration

- Let a source be stationary in the frame S', i.e. it is moving with a speed v in the x-direction in the S frame. Let the wavevector in the S' frame be in the $\mathrm{x}'-y'$ plane, making an angle θ' with the x'-axis. That is, $\tan\theta'=k'_y/k'_x$.

- As seen from frame S, the wavevector makes an angle θ with the x-axis, with

  $$\tan\theta=\frac{k_y}{k_x}=\frac{k'_y}{\gamma k'_x+\gamma\beta\left|\vec{\mathbf{k}}'\right|}=\frac{\sin\theta'}{\gamma(\cos\theta'+\beta)}$$

- Thus the direction of emission of an EM wave from a source changes if the source is moving. This is the phenomenon of aberration.
- Note that $\theta' = 0 \Rightarrow \theta = 0$, $\theta' = \pi \Rightarrow \theta = \pi$, however $\theta' = \pi/2 \Rightarrow \theta = 1/\gamma\beta$. Thus, aberration tends to focus the directions of emitted waves towards the direction of motion of the source.

Doppler Effect

- Equating the coefficients of (ct) on the two sides of the phase-equality, one gets

$$\left|\vec{\mathbf{k}}\right| = \gamma\beta k'_x + \gamma\left|\vec{\mathbf{k}}'\right| = \gamma(1+\beta\cos\theta')\left|\vec{\mathbf{k}}'\right|$$

- Thus, $\omega = \omega'\gamma(1+\beta\cos\theta')$. This change of frequency due to the motion of the source is Doppler effect. I Note the three special cases:

$$\theta' = 0 \Rightarrow \omega = \omega'\sqrt{(1+\beta)/(1-\beta)}$$
$$\theta' = \pi \Rightarrow \omega = \omega'\sqrt{(1-\beta)/(1+\beta)}$$
$$\theta' = \pi/2 \Rightarrow \omega = \omega'\gamma.$$

The first two cases are the blue-shift and red-shift associated with approaching and receding sources, respectively, present even without relativity (though the magnitudes are different). The third is the transverse Doppler shift, which is absent if relativistic effects are ignored.

Intensity of the Wave

- Intensity $\Rightarrow$ power transmitted $\Rightarrow$ Poynting vector.
- The magnitude of the Poynting vector $\vec{\mathbf{N}}$ is

$$\vec{\mathbf{N}} = \left|\vec{\mathbf{E}}\times\vec{\mathbf{H}}\right| = \frac{1}{\mu_0 c}\left|E\right|^2 = \frac{1}{\mu_0 c}\left(\left|\vec{\mathbf{E}}_{\parallel}\right|^2 + \left|\vec{\mathbf{E}}_T\right|^2\right.$$

- We know that

$$\vec{\mathbf{E}}_{\parallel} = \vec{\mathbf{E}}'_{\parallel}, \ \vec{\mathbf{E}}_T = \gamma(\vec{\mathbf{E}}'_T - \vec{\mathbf{v}}\times\vec{\mathbf{B}}'_T)$$

- Let us define $\vec{\mathrm{k}}'$ to be in the x-y plane.
- separate the $\vec{\mathbf{E}}'$ into two components:
    - o Component in the x-y plane
    - o Component normal to the x-y plane

$\vec{\mathbf{k}}$ and $\vec{\mathbf{E}}$ in the x-y plane

- The components of $\vec{\mathbf{k}}', \vec{\mathbf{E}}', \vec{\mathbf{B}}'$ are

$$\vec{\mathbf{k}}' = \left|\vec{\mathbf{k}}'\right| \cos\theta' \hat{\mathrm{x}} + \left|\vec{\mathbf{k}}'\right| \sin\theta' \hat{y}$$

$$\vec{\mathbf{E}}_{\|}' = -\left|\vec{\mathbf{E}}'\right| \sin\theta' \hat{\mathrm{x}} + \left|\vec{\mathbf{E}}'\right| \cos\theta' \hat{y}$$

$$\vec{\mathbf{B}}' = (\left|\vec{\mathbf{E}}'\right| / c)\, \hat{\mathbf{z}}$$

- The Poynting vector becomes

$$\left|\vec{\mathbf{N}}\right| = \frac{1}{\mu_0 c} \left|\vec{\mathbf{E}}'\right|^2 \left[\sin^2\theta' + \gamma^2 \left(\cos\theta' + \frac{v}{c} \right)^2 \right]$$

$$= \left|\vec{\mathbf{N}}'\right| \left[\sin^2\theta' + \gamma^2 \left(\cos\theta' + \frac{v}{c} \right)^2 \right]$$

- Enhancement along the direction of motion of the source

Lorentz Group

In physics and mathematics, the Lorentz group is the group of all Lorentz transformations of Minkowski spacetime, the classical and quantum setting for all (nongravitational) physical phenomena. The Lorentz group is named for the Dutch physicist Hendrik Lorentz.

For example, the following laws, equations, and theories respect Lorentz symmetry:

- The kinematical laws of special relativity
- Maxwell's field equations in the theory of electromagnetism
- The Dirac equation in the theory of the electron
- The Standard model of particle physics

The Lorentz group expresses the fundamental symmetry of space and time of all known fundamental laws of nature. In general relativity physics is that of special relativity in small enough regions of spacetime.

Basic Properties

The Lorentz group is a subgroup of the Poincaré group—the group of all isometries of Minkowski spacetime. Lorentz transformations are, precisely, isometries that leave the origin fixed. Thus, the Lorentz group is an isotropy subgroup of the isometry group of Minkowski spacetime. For this reason, the Lorentz group is sometimes called the homogeneous Lorentz group while the Poincaré group is sometimes called the *inhomogeneous Lorentz group*. Lorentz transformations are examples of linear transformations; general isometries of Minkowski spacetime are affine transformations. Mathematically, the Lorentz group may be described as the generalized orthogonal group O(1,3), the matrix Lie group that preserves the quadratic form

$$(t, x, y, z) \mapsto t^2 - x^2 - y^2 - z^2$$

on R^4. This quadratic form is, when put on matrix form, interpreted in physics as the metric tensor of Minkowski spacetime.

The Lorentz group is a six-dimensional noncompact non-abelian real Lie group that is not connected. The four connected components are not simply connected, but rather doubly connected. The identity component (i.e., the component containing the identity element) of the Lorentz group is itself a group, and is often called the restricted Lorentz group, and is denoted $SO^+(1,3)$. The restricted Lorentz group consists of those Lorentz transformations that preserve the orientation of space and direction of time. The restricted Lorentz group has often been presented through a facility of biquaternion algebra.

The restricted Lorentz group arises in other ways in pure mathematics. For example, it arises as the point symmetry group of a certain ordinary differential equation. This fact also has physical significance.

Connected Components

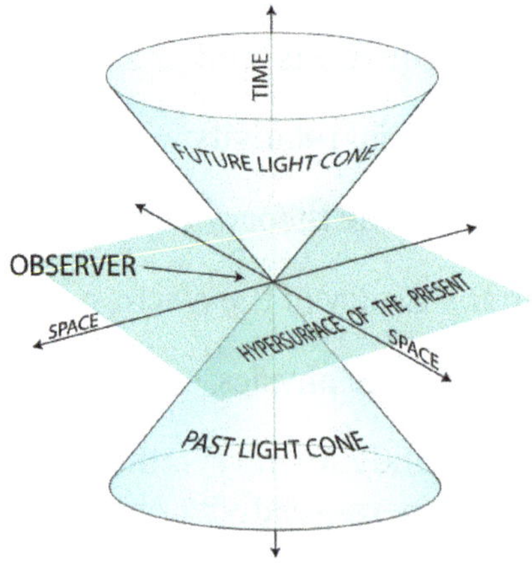

Light cone in 2D space plus a time dimension.

Because it is a Lie group, the Lorentz group O(1,3) is both a group and admits a topological description as a smooth manifold. As a manifold, it has four connected components. Intuitively, this means that it consists of four topologically separated pieces.

The four connected components can be categorized by two transformation properties its elements have:

- some elements are reversed under time-inverting Lorentz transformations, for example, a future-pointing timelike vector would be inverted to a past-pointing vector
- some elements have orientation reversed by improper Lorentz transformations, for example, certain vierbein (tetrads)

Lorentz transformations that preserve the direction of time are called orthochronous. The subgroup of orthochronous transformations is often denoted $O^{+}(1,3)$. Those that preserve orientation are called proper, and as linear transformations they have determinant +1. (The improper Lorentz transformations have determinant −1.) The subgroup of proper Lorentz transformations is denoted SO(1,3).

The subgroup of all Lorentz transformations preserving both orientation and direction of time is called the proper, orthochronous Lorentz group or restricted Lorentz group, and is denoted by $SO^{+}(1, 3)$. (Note that some authors refer to SO(1,3) or even O(1,3) when they actually mean $SO^{+}(1, 3)$.)

The set of the four connected components can be given a group structure as the quotient group $O(1,3)/SO^{+}(1,3)$, which is isomorphic to the Klein four-group. Every element in O(1,3) can be written as the semidirect product of a proper, orthochronous transformation and an element of the discrete group

$$\{1, P, T, PT\}$$

where P and T are the space inversion and time reversal operators:

$$P = \mathrm{diag}(1, -1, -1, -1)$$

$$T = \mathrm{diag}(-1, 1, 1, 1).$$

Thus an arbitrary Lorentz transformation can be specified as a proper, orthochronous Lorentz transformation along with a further two bits of information, which pick out one of the four connected components. This pattern is typical of finite-dimensional Lie groups.

Restricted Lorentz Group

The restricted Lorentz group is the identity component of the Lorentz group, which means that it consists of all Lorentz transformations that can be connected to the identity by a continuous curve lying in the group. The restricted Lorentz group is a connected normal subgroup of the full Lorentz group with the same dimension, in this case with dimension six.

The restricted Lorentz group is generated by ordinary spatial rotations and Lorentz boosts (which can be thought of as hyperbolic rotations in a plane that includes a time-like direction). Since every proper, orthochronous Lorentz transformation can be written as a product of a rotation (specified by 3 real parameters) and a boost (also specified by 3 real parameters), it takes 6 real parameters to specify an arbitrary proper orthochronous Lorentz transformation. This is one way to understand why the restricted Lorentz group is six-dimensional.

The set of all rotations forms a Lie subgroup isomorphic to the ordinary rotation group SO(3). The set of all boosts, however, does *not* form a subgroup, since composing two boosts does not, in general, result in another boost. (Rather, a pair of non-colinear boosts is equivalent to a boost and a rotation, and this relates to Thomas rotation.) A boost in some direction, or a rotation about some axis, generates a one-parameter subgroup.

Surfaces of Transitivity

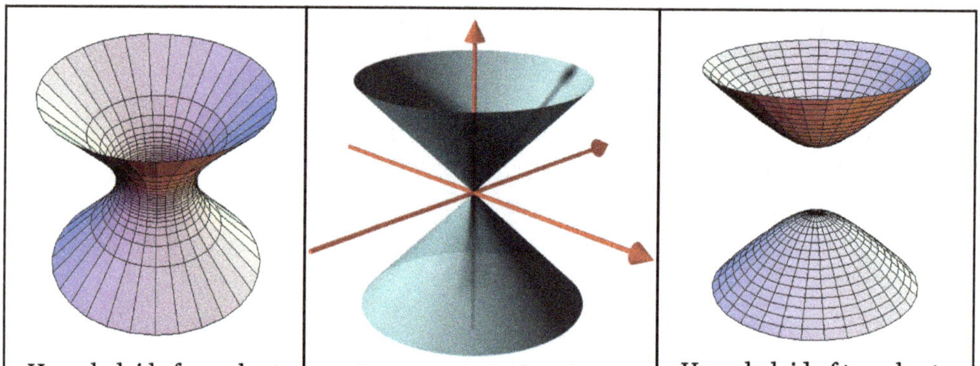

Hyperboloid of one sheet | Common conical surface | Hyperboloid of two sheets

If a group G acts on a space V, then a surface $S \subset V$ is a surface of transitivity if S is invariant under G, i.e., $gs \in S \; \forall g \in G, \forall s \in S$, and for any two points $s_1, s_2 \in S$ there is a $g \in G$ such that $gs_1 = s_2$. By definition of the Lorentz group, it preserves the quadratic form

$$Q(x) = x_0^2 - x_1^2 - x_2^2 - x_3^2.$$

The surfaces of transitivity of the orthochronous Lorentz group $O^+(1, 3)$, $Q(x) = \text{const.}$ of spacetime are the following:

- $Q(x) > 0$, $x_0 > 0$ is the upper branch of a hyperboloid of two sheets.
- $Q(x) > 0$, $x_0 < 0$ is the lower branch of this hyperboloid.

- $Q(x) = 0$, $x_0 > 0$ is the upper branch of the light cone.
- $Q(x) = 0$, $x_0 < 0$ is the lower branch of the light cone.
- $Q(x) < 0$ is a hyperboloid of one sheet.
- The origin $x_0 = x_1 = x_2 = x_3 = 0$.

These surfaces are 3-dimensional, so the images are not faithful, but they are faithful for the corresponding facts about $O^+(1, 2)$. For the full Lorentz group, the surfaces of transitivity are only four since the transformation T takes an upper branch of a hyperboloid (cone) to a lower one and vice versa.

These observations constitute a good starting point for finding all infinite-dimensional unitary representations of the Lorentz group, in fact, of the Poincaré group, using the method of induced representations. One begins with a "standard vector", one for each surface of transitivity, and then ask which subgroup preserves these vectors. These subgroups are called little groups by physicists. The problem is then essentially reduced to the easier problem of finding representations of the little groups. For example, a standard vector in one of the hyperbolas of two sheets could be suitably chosen as $(m, 0, 0, 0)$. For each $m \neq 0$, the vector pierces exactly one sheet. In this case the little group is SO(3), the rotation group, all of whose representations are known. The precise infinite-dimensional unitary representation under which a particle transforms is part of its classification. Not all representations can correspond to physical particles (as far as is known). Standard vectors on the one-sheeted hyperbolas would correspond to tachyons. Particles on the light cone are photons, and more hypothetically, gravitons. The "particle" corresponding to the origin is the vacuum.

Relation to the Möbius Group

The restricted Lorentz group $SO^+(1, 3)$ is isomorphic to the projective special linear group PSL(2,C), which is in turn isomorphic to the Möbius group, the symmetry group of conformal geometry on the Riemann sphere. (This observation was utilized by Roger Penrose as the starting point of twistor theory.)

This may be shown by constructing a surjective homomorphism of Lie groups from SL(2,C) to $SO^+(1,3)$, which we will call the spinor map. This proceeds as follows:

We can define an action of SL(2,C) on Minkowski spacetime by writing a point of spacetime as a two-by-two Hermitian matrix in the form

$$X = \begin{bmatrix} t+z & x-iy \\ x+iy & t-z \end{bmatrix}.$$

This presentation has the pleasant feature that

$$\det X = t^2 - x^2 - y^2 - z^2.$$

Therefore, we have identified the space of Hermitian matrices (which is four-dimensional, as a *real* vector space) with Minkowski spacetime in such a way that the determinant of a Hermitian matrix is the squared length of the corresponding vector in Minkowski spacetime. SL(2,C) acts on the space of Hermitian matrices via

$$X \mapsto PXP^*$$

where P^* is the Hermitian transpose of P, and this action preserves the determinant. Therefore, SL(2,C) acts on Minkowski spacetime by (linear) isometries. This defines a map from SL(2,C) to the Lorentz group $SO^+(1,3)$, and the map is evidently a homomorphism. This is the spinor map.

The kernel of the spinor map is the two element subgroup $\pm I$, and it happens that the map is surjective. By the first isomorphism theorem, the quotient group PSL(2,C) = SL(2,C) / $\{\pm I\}$ is isomorphic to $SO^+(1,3)$.

Appearance of the Night Sky

This isomorphism has the consequence that Möbius transformations of the Riemann sphere represent the way that Lorentz transformations change the appearance of the night sky, as seen by an observer who is maneuvering at relativistic velocities relative to the "fixed stars".

Suppose the "fixed stars" live in Minkowski spacetime and are modeled by points on the celestial sphere. Then a given point on the celestial sphere can be associated with $\xi = u + iv$, a complex number that corresponds to the point on the Riemann sphere, and can be identified with a null vector (a light-like vector) in Minkowski space

$$\begin{bmatrix} u^2+v^2+1 \\ 2u \\ -2v \\ u^2+v^2-1 \end{bmatrix}$$

or the Hermitian matrix

$$N = 2\begin{bmatrix} u^2+v^2 & u+iv \\ u-iv & 1 \end{bmatrix}.$$

The set of real scalar multiples of this null vector, called a *null line* through the origin, represents a *line of sight* from an observer at a particular place and time (an arbitrary event we can identify with the origin of Minkowski spacetime) to various distant objects, such as stars. Then the points of the celestial sphere (equivalently, lines of sight) are identified with certain Hermitian matrices.

Conjugacy Classes

Because the restricted Lorentz group $SO^+(1, 3)$ is isomorphic to the Möbius group PS-L(2,C), its conjugacy classes also fall into five classes:

- Elliptic transformations
- Hyperbolic transformations
- Loxodromic transformations
- Parabolic transformations
- The trivial identity transformation

In the section on Möbius transformations, it is explained how this classification arises by considering the fixed points of Möbius transformations in their action on the Riemann sphere, which corresponds here to null eigenspaces of restricted Lorentz transformations in their action on Minkowski spacetime.

An example of each type is given in the subsections below, along with the effect of the one-parameter subgroup it generates (e.g., on the appearance of the night sky).

The Möbius transformations are the conformal transformations of the Riemann sphere (or celestial sphere). Then conjugating with an arbitrary element of SL(2,C) obtains the following examples of arbitrary elliptic, hyperbolic, loxodromic, and parabolic (restricted) Lorentz transformations, respectively. The effect on the flow lines of the corresponding one-parameter subgroups is to transform the pattern seen in the examples by some conformal transformation. For example, an elliptic Lorentz transformation can have any two distinct fixed points on the celestial sphere, but points still flow along circular arcs from one fixed point toward the other. The other cases are similar.

Elliptic

An elliptic element of SL(2,C) is

$$P_1 = \begin{bmatrix} \exp(i\theta/2) & 0 \\ 0 & \exp(-i\theta/2) \end{bmatrix}$$

and has fixed points $\xi = 0, \infty$. Writing the action as $X \mapsto P_1 X P_1^*$ and collecting terms, the spinor map converts this to the (restricted) Lorentz transformation

$$Q_1 = \begin{bmatrix} 1 & 0 & 0 & 0 \\ 0 & \cos(\theta) & -\sin(\theta) & 0 \\ 0 & \sin(\theta) & \cos(\theta) & 0 \\ 0 & 0 & 0 & 1 \end{bmatrix} = \exp\left(\theta \begin{bmatrix} 0 & 0 & 0 & 0 \\ 0 & 0 & -1 & 0 \\ 0 & 1 & 0 & 0 \\ 0 & 0 & 0 & 0 \end{bmatrix}\right).$$

This transformation then represents a rotation about the z axis, $\exp(i\theta J_z)$. The one-parameter subgroup it generates is obtained by taking θ to be a real variable, the rotation angle, instead of a constant.

The corresponding continuous transformations of the celestial sphere (except for the identity) all share the same two fixed points, the North and South poles. The transformations move all other points around latitude circles so that this group yields a continuous counterclockwise rotation about the z axis as θ increases. The *angle doubling* evident in the spinor map is a characteristic feature of *spinorial double coverings.*

Hyperbolic

A hyperbolic element of SL(2,C) is

$$P_2 = \begin{bmatrix} \exp(\beta/2) & 0 \\ 0 & \exp(-\beta/2) \end{bmatrix}$$

and has fixed points $\xi = 0, \infty$. Under stereographic projection from the Riemann sphere to the Euclidean plane, the effect of this Möbius transformation is a dilation from the origin.

The spinor map converts this to the Lorentz transformation

$$Q_2 = \begin{bmatrix} \cosh(\beta) & 0 & 0 & \sinh(\beta) \\ 0 & 1 & 0 & 0 \\ 0 & 0 & 1 & 0 \\ \sinh(\beta) & 0 & 0 & \cosh(\beta) \end{bmatrix} = \exp\left(\beta \begin{bmatrix} 0 & 0 & 0 & 1 \\ 0 & 0 & 0 & 0 \\ 0 & 0 & 0 & 0 \\ 1 & 0 & 0 & 0 \end{bmatrix} \right).$$

This transformation represents a boost along the z axis with rapidity β. The one-parameter subgroup it generates is obtained by taking β to be a real variable, instead of a constant. The corresponding continuous transformations of the celestial sphere (except for the identity) all share the same fixed points (the North and South poles), and they move all other points along longitudes away from the South pole and toward the North pole.

Loxodromic

A loxodromic element of SL(2,C) is

$$P_3 = P_2 P_1 = P_1 P_2 = \begin{bmatrix} \exp\left((\beta + i\theta)/2\right) & 0 \\ 0 & \exp\left(-(\beta + i\theta)/2\right) \end{bmatrix}$$

and has fixed points $\xi = 0, \infty$. The spinor map converts this to the Lorentz transformation

$$Q_3 = Q_2 Q_1 = Q_1 Q_2.$$

The one-parameter subgroup this generates is obtained by replacing $\beta + i\theta$ with any real multiple of this complex constant. (If β, θ vary independently, then a *two-dimensional* abelian subgroup is obtained, consisting of simultaneous rotations about the z axis and boosts along the z-axis; in contrast, the *one-dimensional* subgroup discussed here consists of those elements of this two-dimensional subgroup such that the rapidity of the boost and angle of the rotation have a *fixed ratio*.)

The corresponding continuous transformations of the celestial sphere (excepting the identity) all share the same two fixed points (the North and South poles). They move all other points away from the South pole and toward the North pole (or vice versa), along a family of curves called loxodromes. Each loxodrome spirals infinitely often around each pole.

Parabolic

A parabolic element of SL(2,C) is

$$P_4 = \begin{bmatrix} 1 & \alpha \\ 0 & 1 \end{bmatrix}$$

and has the single fixed point $\xi = \infty$ on the Riemann sphere. Under stereographic projection, it appears as an ordinary translation along the real axis.

The spinor map converts this to the matrix (representing a Lorentz transformation)

$$Q_4 = \begin{bmatrix} 1+|\alpha|^2/2 & \mathrm{Re}(\alpha) & \mathrm{Im}(\alpha) & -|\alpha|^2/2 \\ \mathrm{Re}(\alpha) & 1 & 0 & -\mathrm{Re}(\alpha) \\ -\mathrm{Im}(\alpha) & 0 & 1 & \mathrm{Im}(\alpha) \\ |\alpha|^2/2 & \mathrm{Re}(\alpha) & \mathrm{Im}(\alpha) & 1-|\alpha|^2/2 \end{bmatrix}$$

$$= \exp \begin{bmatrix} 0 & \mathrm{Re}(\alpha) & \mathrm{Im}(\alpha) & 0 \\ \mathrm{Re}(\alpha) & 0 & 0 & -\mathrm{Re}(\alpha) \\ -\mathrm{Im}(\alpha) & 0 & 0 & \mathrm{Im}(\alpha) \\ 0 & \mathrm{Re}(\alpha) & \mathrm{Im}(\alpha) & 0 \end{bmatrix}.$$

This generates a two-parameter abelian subgroup, which is obtained by considering α

a complex variable rather than a constant. The corresponding continuous transformations of the celestial sphere (except for the identity transformation) move points along a family of circles that are all tangent at the North pole to a certain great circle. All points other than the North pole itself move along these circles.

Parabolic Lorentz transformations are often called null rotations, since they preserve null vectors, just as rotations preserve timelike vectors and boosts preserve spacelike vectors. Since these are likely to be the least familiar of the four types of nonidentity Lorentz transformations (elliptic, hyperbolic, loxodromic, parabolic), it is illustrated here how to determine the effect of an example of a parabolic Lorentz transformation on Minkowski spacetime.

The matrix given above yields the transformation

$$\begin{bmatrix} t \\ x \\ y \\ z \end{bmatrix} \to \begin{bmatrix} t \\ x \\ y \\ z \end{bmatrix} + \mathrm{Re}(\alpha)\begin{bmatrix} x \\ t-z \\ 0 \\ x \end{bmatrix} + \mathrm{Im}(\alpha)\begin{bmatrix} y \\ 0 \\ z-t \\ y \end{bmatrix} + \frac{|\alpha|^2}{2}\begin{bmatrix} t-z \\ 0 \\ 0 \\ t-z \end{bmatrix}.$$

Now, without loss of generality, pick $Im(a)=0$. Differentiating this transformation with respect to the now real group parameter a and evaluating at $a=0$ produces the corresponding vector field (first order linear partial differential operator),

$$x(\partial_t + \partial_z) + (t-z)\partial_x.$$

Apply this to a function $f(t,x,y,z)$, and demand that it stays invariant, i.e., it is annihilated by this transformation. The solution of the resulting first order linear partial differential equation can be expressed in the form

$$f(t,x,y,z) = F(y, t-z, t^2 - x^2 - z^2),$$

where F is an *arbitrary* smooth function. The arguments of F give three *rational invariants* describing how points (events) move under this parabolic transformation, as they themselves do not move,

$$y = c_1, \quad t - z = c_2, \quad t^2 - x^2 - z^2 = c_3.$$

Choosing real values for the constants on the right hand sides yields three conditions, and thus specifies a curve in Minkowski spacetime. This curve is an orbit of the transformation.

The form of the rational invariants shows that these flowlines (orbits) have a simple description: suppressing the inessential coordinate y, each orbit is the intersection of a *null plane*, $t = z+c_2$, with a *hyperboloid*, $t^2-x^2-z^2 = c_3$. The case $c_3 = 0$ has the hyperboloid degenerate to a light cone with the orbits becoming parabolas lying in corresponding null planes.

A particular null line lying on the light cone is left *invariant*; this corresponds to the unique (double) fixed point on the Riemann sphere mentioned above. The other null lines through the origin are "swung around the cone" by the transformation. Following the motion of one such null line as α increases corresponds to following the motion of a point along one of the circular flow lines on the celestial sphere, as described above.

A choice $Re(\alpha)=0$ instead, produces similar orbits, now with the roles of x and y interchanged.

Parabolic transformations lead to the gauge symmetry of massless particles (like photons) with helicity $|h| \geq 1$. In the above explicit example, a massless particle moving in the z direction, so with 4-momentum $P=(p,0,0,p)$, is not affected at all by the x-boost and y-rotation combination K_x-J_y displayed above, in the "little group" of its motion. This is evident from the explicit transformation law discussed: like any light-like vector, P itself is now invariant, i.e., all traces or effects of α have disappeared. $c_1 = c_2 = c_3 = 0$, in the special case discussed. (The other similar generator, K_y+J_x as well as it and J_z comprise altogether the little group of the lightlike vector, isomorphic to $E(2)$.)

Lie Algebra

As with any Lie group, the best way to study many aspects of the Lorentz group is via its Lie algebra. The Lorentz group is a subgroup of the diffeomorphism group of R^4 and therefore its Lie algebra can be identified with vector fields on R^4. In particular, the vectors that generate isometries on a space are its Killing vectors, which provides a convenient alternative to the left-invariant vector field for calculating the Lie algebra. We can write down a set of six generators:

- vector fields on R^4 generating three rotations $i\,J$,

$$-y\partial_x + x\partial_y \equiv iJ_z\,, \qquad -z\partial_y + y\partial_z \equiv iJ_x\,, \qquad -x\partial_z + z\partial_x \equiv iJ_y\,;$$

- vector fields on R^4 generating three boosts $i\,K$,

$$x\partial_t + t\partial_x \equiv iK_x\,, \qquad y\partial_t + t\partial_y \equiv iK_y\,, \qquad z\partial_t + t\partial_z \equiv iK_z.$$

It may be helpful to briefly recall here how to obtain a one-parameter group from a vector field, written in the form of a first order linear partial differential operator such as

$$-y\partial_x + x\partial_y.$$

The corresponding initial value problem is

$$\frac{\partial x}{\partial \lambda} = -y, \frac{\partial y}{\partial \lambda} = x, x(0) = x_0, y(0) = y_0.$$

The solution can be written

$$x(\lambda) = x_0 \cos(\lambda) - y_0 \sin(\lambda), y(\lambda) = x_0 \sin(\lambda) + y_0 \cos(\lambda)$$

or

$$\begin{bmatrix} t \\ x \\ y \\ z \end{bmatrix} = \begin{bmatrix} 1 & 0 & 0 & 0 \\ 0 & \cos(\lambda) & -\sin(\lambda) & 0 \\ 0 & \sin(\lambda) & \cos(\lambda) & 0 \\ 0 & 0 & 0 & 1 \end{bmatrix} \begin{bmatrix} t_0 \\ x_0 \\ y_0 \\ z_0 \end{bmatrix}$$

where we easily recognize the one-parameter matrix group of rotations $\exp(i\lambda J_z)$ about the z axis. Differentiating with respect to the group parameter λ and setting it λ=0 in that result, we recover the standard matrix,

$$iJ_z = \begin{bmatrix} 0 & 0 & 0 & 0 \\ 0 & 0 & -1 & 0 \\ 0 & 1 & 0 & 0 \\ 0 & 0 & 0 & 0 \end{bmatrix},$$

which corresponds to the vector field we started with. This illustrates how to pass between matrix and vector field representations of elements of the Lie algebra.

Reversing the procedure we see that the Möbius transformations that Dcorrespond to our six generators arise from exponentiating respectively $\beta/2$ (for the three boosts) or $i\theta/2$ (for the three rotations) times the three Pauli matrices

$$\sigma_1 = \begin{bmatrix} 0 & 1 \\ 1 & 0 \end{bmatrix}, \sigma_2 = \begin{bmatrix} 0 & -i \\ i & 0 \end{bmatrix}, \sigma_3 = \begin{bmatrix} 1 & 0 \\ 0 & -1 \end{bmatrix}.$$

For our purposes, another generating set is more convenient. The following table lists the six generators, in which

- The first column gives a generator of the flow under the Möbius action (after stereographic projection from the Riemann sphere) as a *real* vector field on the Euclidean plane.

- The second column gives the corresponding one-parameter subgroup of Möbius transformations.
- The third column gives the corresponding one-parameter subgroup of Lorentz transformations (the image under our homomorphism of preceding one-parameter subgroup).
- The fourth column gives the corresponding generator of the flow under the Lorentz action as a real vector field on Minkowski spacetime.

Notice that the generators consist of

- Two parabolics (null rotations)
- One hyperbolic (boost in the ∂_z direction)
- Three elliptics (rotations about the *x,y,z* axes, respectively)

Vector field on R^2	**One-parameter subgroup of SL(2,C), representing Möbius transformations**	**One-parameter subgroup of $SO^+(1,3)$, representing Lorentz transformations**	**Vector field on R^4**
		Parabolic	
∂_u	$\begin{bmatrix} 1 & \alpha \\ 0 & 1 \end{bmatrix}$	$\begin{bmatrix} 1+\alpha^2/2 & \alpha & 0 & -\alpha^2/2 \\ \alpha & 1 & 0 & -\alpha \\ 0 & 0 & 1 & 0 \\ \alpha^2/2 & \alpha & 0 & 1-\alpha^2/2 \end{bmatrix}$	$X_1 =$ $x(\partial_t+\partial_z)+(t-z)\partial_x$
∂_v	$\begin{bmatrix} 1 & i\alpha \\ 0 & 1 \end{bmatrix}$	$\begin{bmatrix} 1+\alpha^2/2 & 0 & \alpha & -\alpha^2/2 \\ 0 & 1 & 0 & 0 \\ \alpha & 0 & 1 & -\alpha \\ \alpha^2/2 & 0 & \alpha & 1-\alpha^2/2 \end{bmatrix}$	$X_2 =$ $y(\partial_t+\partial_z)+(t-z)\partial_y$
		Hyperbolic	
$\frac{1}{2}(u\partial_u+v\partial_v)$	$\begin{bmatrix} \exp\left(\frac{\beta}{2}\right) & 0 \\ 0 & \exp\left(-\frac{\beta}{2}\right) \end{bmatrix}$	$\begin{bmatrix} \cosh(\beta) & 0 & 0 & \sinh(\beta) \\ 0 & 1 & 0 & 0 \\ 0 & 0 & 1 & 0 \\ \sinh(\beta) & 0 & 0 & \cosh(\beta) \end{bmatrix}$	$X_3 =$ $z\partial_t+t\partial_z$

Elliptic			
$\frac{1}{2}\left(-v\partial_u + u\partial_v\right)$	$\begin{bmatrix} \exp\left(\frac{i\theta}{2}\right) & 0 \\ 0 & \exp\left(\frac{-i\theta}{2}\right) \end{bmatrix}$	$\begin{bmatrix} 1 & 0 & 0 & 0 \\ 0 & \cos(\theta) & -\sin(\theta) & 0 \\ 0 & \sin(\theta) & \cos(\theta) & 0 \\ 0 & 0 & 0 & 1 \end{bmatrix}$	$X_4 = -y\partial_x + x\partial_y$
$\frac{v^2 - u^2 - 1}{2}\partial_u - uv\partial_v$	$\begin{bmatrix} \cos\left(\frac{\theta}{2}\right) & -\sin\left(\frac{\theta}{2}\right) \\ \sin\left(\frac{\theta}{2}\right) & \cos\left(\frac{\theta}{2}\right) \end{bmatrix}$	$\begin{bmatrix} 1 & 0 & 0 & 0 \\ 0 & \cos(\theta) & 0 & \sin(\theta) \\ 0 & 0 & 1 & 0 \\ 0 & -\sin(\theta) & 0 & \cos(\theta) \end{bmatrix}$	$X_5 = -x\partial_z + z\partial_x$
$uv\partial_u + \frac{1-u^2+v^2}{2}\partial_v$	$\begin{bmatrix} \cos\left(\frac{\theta}{2}\right) & i\sin\left(\frac{\theta}{2}\right) \\ i\sin\left(\frac{\theta}{2}\right) & \cos\left(\frac{\theta}{2}\right) \end{bmatrix}$	$\begin{bmatrix} 1 & 0 & 0 & 0 \\ 0 & 1 & 0 & 0 \\ 0 & 0 & \cos(\theta) & -\sin(\theta) \\ 0 & 0 & \sin(\theta) & \cos(\theta) \end{bmatrix}$	$X_6 = -z\partial_y + y\partial_z$

Let's verify one line in this table. Start with

$$\sigma_2 = \begin{bmatrix} 0 & i \\ -i & 0 \end{bmatrix}.$$

Exponentiate:

$$\exp\left(\frac{i\theta}{2}\sigma_2\right) = \begin{bmatrix} \cos(\theta/2) & -\sin(\theta/2) \\ \sin(\theta/2) & \cos(\theta/2) \end{bmatrix}.$$

This element of SL(2,C) represents the one-parameter subgroup of (elliptic) Möbius transformations:

$$\xi \mapsto \frac{\cos(\theta/2)\xi - \sin(\theta/2)}{\sin(\theta/2)\xi + \cos(\theta/2)}.$$

Next,

$$\frac{d\xi}{d\theta}\Big|_{\theta=0} = -\frac{1+\xi^2}{2}.$$

The corresponding vector field on C (thought of as the image of S^2 under stereographic projection) is

$$-\frac{1+\xi^2}{2}\partial_\xi.$$

Writing $\xi = u + iv$, this becomes the vector field on R²

$$-\frac{1+u^2-v^2}{2}\partial_u - uv\partial_v.$$

Returning to our element of SL(2,*C*), writing out the action $X \mapsto PXP^*$ and collecting terms, we find that the image under the spinor map is the element of SO⁺(1,3)

$$\begin{bmatrix} 1 & 0 & 0 & 0 \\ 0 & \cos(\theta) & 0 & \sin(\theta) \\ 0 & 0 & 1 & 0 \\ 0 & -\sin(\theta) & 0 & \cos(\theta) \end{bmatrix}.$$

Differentiating with respect to θ at θ=0, yields the corresponding vector field on R⁴,

$$z\partial_x - x\partial_z.$$

This is evidently the generator of counterclockwise rotation about the *y* axis.

Subgroups of the Lorentz Group

The subalgebras of the Lie algebra of the Lorentz group can be enumerated, up to conjugacy, from which we can list the closed subgroups of the restricted Lorentz group, up to conjugacy. We can readily express the result in terms of the generating set given in the table.

The one-dimensional subalgebras of course correspond to the four conjugacy classes of elements of the Lorentz group:

- X_1 generates a one-parameter subalgebra of parabolics SO(0,1),
- X_3 generates a one-parameter subalgebra of boosts SO(1,1),
- X_4 generates a one-parameter of rotations SO(2),
- $X_3 + aX_4$ (for any $a \neq 0$) generates a one-parameter subalgebra of loxodromic transformations.

(Strictly speaking the last corresponds to infinitely many classes, since distinct a give different classes.) The two-dimensional subalgebras are:

- X_1, X_2 generate an abelian subalgebra consisting entirely of parabolics,

- X_1, X_3 generate a nonabelian subalgebra isomorphic to the Lie algebra of the affine group A(1),
- X_3, X_4 generate an abelian subalgebra consisting of boosts, rotations, and loxodromics all sharing the same pair of fixed points.

The three-dimensional subalgebras are:

- X_1, X_2, X_3 generate a Bianchi V subalgebra, isomorphic to the Lie algebra of Hom(2), the group of *euclidean homotheties*,
- X_1, X_2, X_4 generate a Bianchi VII_0 subalgebra, isomorphic to the Lie algebra of E(2), the euclidean group,
- $X_2, X_2, X_3 + aX_4$, where $a \neq 0$, generate a Bianchi VII_a subalgebra,
- X_1, X_3, X_5 generate a Bianchi VIII subalgebra, isomorphic to the Lie algebra of SL(2,R), the group of isometries of the hyperbolic plane,
- X_4, X_5, X_6 generate a Bianchi IX subalgebra, isomorphic to the Lie algebra of SO(3), the rotation group.

(Here, the Bianchi types refer to the classification of three-dimensional Lie algebras by the Italian mathematician Luigi Bianchi.) The four-dimensional subalgebras are all conjugate to

- X_1, X_2, X_3, X_4 generate a subalgebra isomorphic to the Lie algebra of Sim(2), the group of Euclidean similitudes.

The subalgebras form a lattice, and each subalgebra generates by exponentiation a closed subgroup of the restricted Lie group. From these, all subgroups of the Lorentz group can be constructed, up to conjugation, by multiplying by one of the elements of the Klein four-group.

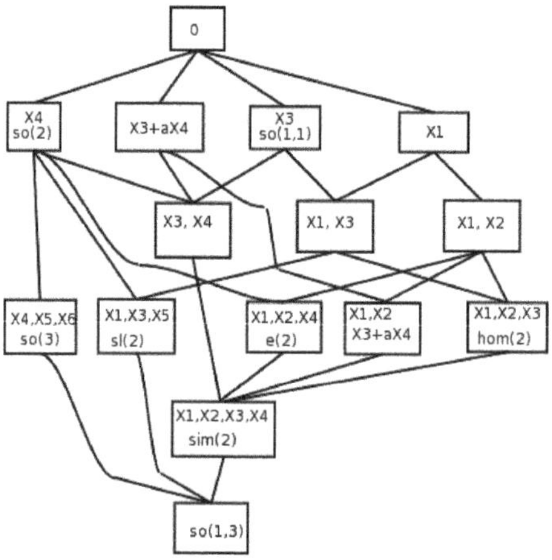

The lattice of subalgebras of the Lie algebra SO(1,3), up to conjugacy.

As with any connected Lie group, the coset spaces of the closed subgroups of the restricted Lorentz group, or homogeneous spaces, have considerable mathematical interest. A few, brief descriptions:

- The group Sim(2) is the stabilizer of a *null line*, i.e., of a point on the Riemann sphere—so the homogeneous space $SO^+(1,3)/Sim(2)$ is the Kleinian geometry that represents conformal geometry on the sphere $S.^2$
- The (identity component of the) Euclidean group SE(2) is the stabilizer of a null vector, so the homogeneous space $SO^+(1,3)/SE(2)$ is the momentum space of a massless particle; geometrically, this Kleinian geometry represents the *degenerate* geometry of the light cone in Minkowski spacetime.
- The rotation group SO(3) is the stabilizer of a timelike vector, so the homogeneous space $SO^+(1,3)/SO(3)$ is the momentum space of a massive particle; geometrically, this space is none other than three-dimensional hyperbolic space H^3.

Covering Groups

We constructed a homomorphism $SL(2, C) \to SO^+(1, 3)$, which we called the spinor map. Since SL(2,C) is simply connected, it is the covering group of the restricted Lorentz group $SO^+(1, 3)$. By restriction we obtain a homomorphism $SU(2) \to SO(3)$. Here, the special unitary group SU(2), which is isomorphic to the group of unit norm quaternions, is also simply connected, so it is the covering group of the rotation group SO(3). Each of these covering maps are twofold covers in the sense that precisely two elements of the covering group map to each element of the quotient. One often says that the restricted Lorentz group and the rotation group are doubly connected. This means that the fundamental group of the each group is isomorphic to the two-element cyclic group Z_2.

(In applications to quantum mechanics, the special linear group SL(2, C) is sometimes called the Lorentz group.)

Twofold coverings are characteristic of spin groups. Indeed, in addition to the double coverings

$Spin^+(1, 3) = SL(2, C) \to SO^+(1, 3)$

$Spin(3) = SU(2) \to SO(3)$

we have the double coverings

$Pin(1, 3) \to O(1, 3)$

$Spin(1, 3) \to SO(1, 3)$

$Spin^+(1, 2) = SU(1, 1) \to SO(1, 2)$

These spinorial double coverings are all closely related to Clifford algebras.

Topology

The left and right groups in the double covering

SU(2) → SO(3)

are deformation retracts of the left and right groups, respectively, in the double covering

SL(2,C) → $SO^+(1,3)$.

But the homogeneous space $SO^+(1,3)/SO(3)$ is homeomorphic to hyperbolic 3-space H^3, so we have exhibited the restricted Lorentz group as a principal fiber bundle with fibers SO(3) and base H^3. Since the latter is homeomorphic to R^3, while SO(3) is homeomorphic to three-dimensional real projective space RP^3, we see that the restricted Lorentz group is *locally* homeomorphic to the product of RP^3 with R^3. Since the base space is contractible, this can be extended to a global homeomorphism.

Generalization to Higher Dimensions

The concept of the Lorentz group has a natural generalization to spacetime of any number of dimensions. Mathematically, the Lorentz group of n+1-dimensional Minkowski space is the group $O(n,1)$ (or $O(1,n)$) of linear transformations of R^{n+1} that preserves the quadratic form

$$(x_1, x_2, \ldots, x_n, x_{n+1}) \mapsto x_1^2 + x_2^2 + \cdots + x_n^2 - x_{n+1}^2.$$

Many of the properties of the Lorentz group in four dimensions (where $n = 3$) generalize straightforwardly to arbitrary n. For instance, the Lorentz group $O(n,1)$ has four connected components, and it acts by conformal transformations on the celestial $(n-1)$-sphere in n+1-dimensional Minkowski space. The identity component $SO^+(n,1)$ is an $SO(n)$-bundle over hyperbolic n-space H^n.

The low-dimensional cases $n = 1$ and $n = 2$ are often useful as "toy models" for the physical case $n = 3$, while higher-dimensional Lorentz groups are used in physical theories such as string theory that posit the existence of hidden dimensions. The Lorentz group $O(n,1)$ is also the isometry group of n-dimensional de Sitter space dS_n, which may be realized as the homogeneous space $O(n,1)/O(n-1,1)$. In particular $O(4,1)$ is the isometry group of the de Sitter universe dS_4, a cosmological model.

Contravariant and Covariant 4-vectors

Qualifications for Being Termed a 4-vector

- A 4-vector is a 4-component object whose components transform under a change of frame either like the differentials , (cdt, dx, dy, dz) or

like the derivatives $\left(\frac{\partial}{\partial(ct)}, \frac{\partial}{\partial x}, \frac{\partial}{\partial y}, \frac{\partial}{\partial z}\right)$.

In the former case, it is a contravariant vector.

In the latter case, it is a covariant vector.

- We shall later see that a vector itself may be treated as a geometrical object, which may be represented by its covariant or contravariant components. We shall represent the 4-vectors by sans-serif notation X. Their contravariant components will be represented by X^k and the covariant components will be represented by $X_k\ (k \in \{0,1,2,3\})$.

Lorentz Transformations of dx

- We already know how the coordinates transform under a boost along x-axis:

$$\begin{pmatrix} cdt' \\ dx' \\ dy' \\ dz' \end{pmatrix} = \begin{pmatrix} \gamma & -\gamma\beta & 0 & 0 \\ -\gamma\beta & \gamma & 0 & 0 \\ 0 & 0 & 1 & 0 \\ 0 & 0 & 0 & 1 \end{pmatrix} \begin{pmatrix} cdt \\ dx \\ dy \\ dz \end{pmatrix}$$

- We write the above equation as

$$dx' = \wedge dx, \qquad or \qquad dx'^k = \sum_m \wedge^k_m dx^m$$

where

$$dx^m = (dx^0, dx^1, dx^2, dx^3) = (cdt, dx, dy, dz)$$

- The above equation may be looked upon simply as a matrix equation at the moment. (Later, we'll interpret it as a tensor equation, but that is getting ahead of ourselves.)

Contravariant Components

- Chain rule for differentials implies

$$dx'^k = \sum \frac{\partial x'^k}{\partial x^m} dx^m$$

Thus, $\wedge^k_m = \partial x'^k / \partial x^m$

- Any 4-component object X that transforms as $X' = \wedge X$ is a contravariant vector. Thus, a contravariant vector transforms as

$$X'^k = \frac{\partial x'^k}{\partial x^m} X^m ,$$

where we have started using the convention where repeated indices are summed over.

A Few Comments

- The definition of $\wedge$ as $\wedge^k_m = \partial x'^k / \partial x^m$ is a general one, valid in all circumstances.
- Here we have considered only Lorentz boosts along x direction, so the 4×4 matrix that appears in Eq. (1) is a special case of $\wedge$.
- However, once we know how a 4-component object behaves under this special Lorentz boost, we can combine this with our prior information on how the "space" components of this object behave under rotations, to figure out how the whole 4-component object behaves under a general Lorentz boost.
- The space components of all the relevant objects we'll consider here are 3-vectors. Hence we know that they behave "as a vector should" under space rotations, and it only remains to check that they transform properly under Lorentz boosts. We shall hence only focus on this last point.

Lorentz Transformations of ∂

- Let us represent

$$\partial_m \equiv (\partial_0, \partial_1, \partial_2, \partial_3) \equiv \frac{\partial}{\partial x^m} = \left(\frac{\partial}{\partial(ct)}, \frac{\partial}{\partial x}, \frac{\partial}{\partial y}, \frac{\partial}{\partial z} \right)$$

- The chain rule for derivatives gives

$$\partial'_k = \frac{\partial}{\partial x'^k} = \frac{\partial x^m}{\partial x'^k} \frac{\partial}{\partial x^m} = \frac{\partial x^m}{\partial x'^k} \partial_m = \overline{\Lambda}^m{}_k \partial_m .$$

- For the boost in x direction with speed v, one gets

$$\overline{\Lambda}^m{}_k = \begin{pmatrix} \gamma & \gamma\beta & 0 & 0 \\ \gamma\beta & \gamma & 0 & 0 \\ 0 & 0 & 1 & 0 \\ 0 & 0 & 0 & 1 \end{pmatrix}.$$

Covariant Vectors

- The requirement in equation may be written as a matrix equation

$$\partial' = \partial\bar{\Lambda},$$

where @ should be written as a row vector.

- Any 4-component object that transforms as

$$Y'_k = \frac{\partial x^m}{\partial x'^k} Y_m = \bar{\Lambda}^m{}_k Y_m$$

is a covariant vector.

Relationship Between Λ and $\bar{\Lambda}$

- Note that

$$\Lambda^m{}_n \bar{\Lambda}^n{}_k = \frac{\partial x'^m}{\partial x^n}\frac{\partial x^n}{\partial x'^k} = \delta^m_k.$$

Therefore, $\Lambda\bar{\Lambda} = 1$ and $\bar{\Lambda} = \Lambda^{-1}$.

- This also leads to

$$X'^m Y'_m = \Lambda^m_n X^n \bar{\Lambda}^k{}_m Y_k = \delta^k_n X^n Y_k = X^k Y_k$$

So the Lorentz transformations do not affect the "inner product" $X^k Y_k$ of a contravariant and a covariant vector. Such a quantity is therefore frame-independent.

Examples of 4-vectors x

The position 4-vector: x, ∂, p, J, A, u, a

- Clearly the position vector

$$x^k = (x^0, x^1, x^2, x^3) = (ct, x, y, z) = (ct, \vec{\mathbf{x}})$$

transforms like dx^k, i.e. $x'^k = \Lambda^k_m x^m$. It is therefore a contravariant 4-vector.

- However also note that

$$x_k = (x_0, x_1, x_2, x_3) = (ct, -x, -y, -z) = (ct, -\vec{\mathbf{x}})$$

transforms like $x'_k = \bar{\Lambda}^m{}_k x_m$. It is therefore a covariant vector.

- From the above, x^k *and* x_k may be interpreted as the contravariant and covariant components of the same object, a 4-vector x.

The Differential Operator ∂ as a 4-vector

- The differential operator

$$\partial_m = (\partial_0, \partial_1, \partial_2, \partial_3) = \left(\frac{\partial}{\partial(ct)}, \frac{\partial}{\partial x}, \frac{\partial}{\partial y}, \frac{\partial}{\partial z} \right) = \left(\frac{\partial}{\partial(ct)}, \nabla \right)$$

is a covariant 4-vector by definition.

- Also note that

$$\partial^m = (\partial^0, \partial^1, \partial^2, \partial^3) = \left(\frac{\partial}{\partial(ct)}, -\frac{\partial}{\partial x}, -\frac{\partial}{\partial y}, -\frac{\partial}{\partial z} \right) = \left(\frac{\partial}{\partial(ct)}, -\nabla \right)$$

transforms like $\partial'^m = \Lambda^m_k \partial^k$. It is therefore a contravariant vector.

- Thus, ∂ may ∂_m be interpreted as the contravariant and covariant components of the same object, a 4-vector ∂.

The Momentum 4-vector p

- We have seen that in relativity, the momentum of a particle travelling with velocity $\vec{\mathbf{u}}$ is $\vec{\mathbf{p}} = m\gamma\vec{\mathbf{u}}$ while its energy is $E = m\gamma c^2$.
- Consider the 4-component quantity $(E/c, \vec{\mathbf{p}})$. In the rest frame of the particle, it equals $(mc, \vec{0})$. Let this be the frame S.
- In the frame S' (moving with a speed v along x direction, as seen from S), the above quantity should be $(m\gamma c, -m\gamma\vec{\mathbf{v}})$ This is obtained by a trasformation through the matrix Λ, as can be checked explicitly. Therefore,

$$p^m = (p^0, p^1, p^2, p^3) = (E/c, \vec{\mathbf{p}})$$

is a contravariant 4-vector.

- Similarly,

$$p_m = (p_0, p_1, p_2, p_3) = (E/c, -\vec{\mathbf{p}})$$

is a covariant 4-vector.

The Current 4-vector J

- Consider a wire of constant cross section, and uniform charge density ρ in frame S, in which no current is flowing, i.e. $\vec{\mathbf{J}}=0$. Let the wire be along x direction.
- In a frame S' moving with a speed v along x direction, one sees a charge density $\rho' = \gamma\rho$ due to Lorentz contraction, and a current density $\vec{\mathbf{J}}' = -\gamma\rho\vec{\mathbf{v}}$.
- Thus, the transformation of the object $(\rho c, \vec{\mathbf{J}})$ to a moving frame can be obtained through the matrix Λ
- Therefore,

$$\mathrm{J}^m = (\mathrm{J}^0, \mathrm{J}^1, \mathrm{J}^2, \mathrm{J}^3) = (\rho c, \vec{\mathbf{J}})$$

is a contravariant 4-vector, while

$$\mathrm{J}_m = (\mathrm{J}_0, \mathrm{J}_1, \mathrm{J}_2, \mathrm{J}_3) = (\rho c, -\vec{\mathbf{J}})$$

is a covariant 4-vector.

Electromagnetic Potential 4-vector A

- In the Lorentz gauge

$$\nabla . \vec{\mathbf{A}} + \frac{1}{c^2}\frac{\partial \phi}{\partial t} = 0,$$

the vector and scalar potentials $\vec{\mathbf{A}}$ and ϕ satisfy the wave equations

$$\frac{\partial^2 \vec{\mathbf{A}}}{\partial (ct)^2} - \nabla^2 \vec{\mathbf{A}} = \mu_0 \vec{\mathbf{J}}, \quad \frac{\partial^2 \phi}{\partial (ct)^2} - \nabla^2 \phi = \frac{\rho}{\varepsilon_0}.$$

- Since the wave equations do not change under Lorentz transformations as seen earlier, clearly $(\phi/c, \vec{\mathbf{A}})$ should have the same transformation properties like $(\rho c, \vec{\mathbf{J}})$
- Therefore,

$$A^m = (A^0, A^1, A^2, A^3) = (\phi/c, \vec{\mathbf{A}})$$

is a contravariant 4-vector, while

$$A_m = (A_0, A_1, A_2, A_3) = (\phi/c, -\vec{\mathbf{A}})$$

is a covariant 4-vector.

4-velocity u

- While defining new 4-vectors in terms of the known ones, one should ensure that they obey the required transformations. The covariant / contravariant indexing (subscripts / superscripts) and the summation convention are useful tools to take care of this.

- The 4-velocity is defined as

$$u^m = c\frac{dx^m}{ds}$$

which is clearly a 4-vector, since dx^m is a 4-vector and

$ds = \sqrt{(c\,dt)^2 - (dx)^2 - (dy)^2 - (dz)^2}$ is a Lorentz-invariant scalar.

- Since $ds = dt/\gamma$ the 4-velocity may be written as

$$u^m = (\gamma c, \gamma\vec{v})$$

- Note that

$$u^m u_m = c^2 \frac{dx^m dx_m}{(ds)^2} = c^2$$

Thus, u is a "unit" 4-vector (in the units c = 1).

4-acceleration a

- The acceleration 4-vector is naturally defined as

$$a^m = c^2 \frac{d^2x^m}{ds^2} = c = \frac{du^m}{ds}$$

- Given that $u^m u_m = c^2$ a constant, one gets $a^m u_m = 0$ Thus, 4-velocity and 4-acceleration are orthogonal to each other (in the 4-dimensional sense): $u.a = 0$.

References

- Feynman, Richard Phillips; Leighton, Robert B.; Sands, Matthew L. (2006). The Feynman lectures on physics (3 vol.). Pearson / Addison-Wesley. ISBN 0-8053-9047-2. : volume 2

- Hatcher, Allen (2002). Algebraic topology. Cambridge: Cambridge University Press. ISBN 0-521-79540-0. See also the "online version". Retrieved July 3, 2005. See Section 1.3 for a beautifully illustrated discussion of covering spaces

- Serway, Raymond A.; Jewett, John W., Jr. (2004). Physics for scientists and engineers, with modern physics. Belmont, [CA.]: Thomson Brooks/Cole. ISBN 0-534-40846-X

- Wigner, E. P. (1939), "On unitary representations of the inhomogeneous Lorentz group", Annals of Mathematics, 40 (1): 149–204, Bibcode:1939AnMat..40..149W, MR 1503456, doi:10.2307/1968551
- Frankel, Theodore (2004). The Geometry of Physics (2nd Ed.). Cambridge: Cambridge University Press. ISBN 0-521-53927-7. An excellent resource for Lie theory, fiber bundles, spinorial coverings, and many other topics

4

Momentum in Special Relativity

Special relativity establishes the relation between space and time. It forms the precise mod-el of motion at any speed. Special relativity provides formulas of the change that occurs in electromagnetic objects under Lorentz transformation as well as the relationship between electricity and magnetism. The aim of this section is to explore the special relativity and electromagnetism. These topics are crucial for a complete understanding of the subject.

Special Relativity

In physics, special relativity (SR, also known as the special theory of relativity or STR) is the generally accepted and experimentally well-confirmed physical theory regarding the relationship between space and time. In Albert Einstein's original pedagogical treatment, it is based on two postulates:

1. The laws of physics are invariant (i.e. identical) in all inertial systems (non-accelerating frames of reference).
2. The speed of light in a vacuum is the same for all observers, regardless of the motion of the light source.

It was originally proposed in 1905 by Albert Einstein in the paper "On the Electrodynamics of Moving Bodies". The inconsistency of Newtonian mechanics with Maxwell's equations of electromagnetism and the lack of experimental confirmation for a hypothesized luminiferous aether led to the development of special relativity, which corrects mechanics to handle situations involving motions at a significant fraction of the speed of light (known as *relativistic velocities*). As of today, special relativity is the most accurate model of motion at any speed. Even so, the Newtonian mechanics model is still useful (due to its simplicity and high accuracy) as an approximation at small velocities relative to the speed of light.

Not until Einstein developed general relativity, to incorporate general (or accelerated) frames of reference and gravity, was the phrase "special relativity" employed. A translation that has often been used is "restricted relativity"; "special" really means "special case".

Special relativity implies a wide range of consequences, which have been experimentally verified, including length contraction, time dilation, relativistic mass, mass–energy equivalence, a universal speed limit and relativity of simultaneity. It has replaced the conventional notion of an absolute universal time with the notion of a time that is dependent on reference frame and spatial position. Rather than an invariant time interval between two events, there is an invariant spacetime interval. Combined with other laws of physics, the two postulates of special relativity predict the equivalence of mass and energy, as expressed in the mass–energy equivalence formula $E = mc^2$, where c is the speed of light in a vacuum.

A defining feature of special relativity is the replacement of the Galilean transformations of Newtonian mechanics with the Lorentz transformations. Time and space cannot be defined separately from each other. Rather space and time are interwoven into a single continuum known as spacetime. Events that occur at the same time for one observer can occur at different times for another.

The theory is "special" in that it only applies in the special case where the curvature of spacetime due to gravity is negligible. In order to include gravity, Einstein formulated general relativity in 1915. Special relativity, contrary to some outdated descriptions, is capable of handling accelerations as well as accelerated frames of reference.

As Galilean relativity is now considered an approximation of special relativity that is valid for low speeds, special relativity is considered an approximation of general relativity that is valid for weak gravitational fields, i.e. at a sufficiently small scale and in conditions of free fall. Whereas general relativity incorporates noneuclidean geometry in order to represent gravitational effects as the geometric curvature of spacetime, special relativity is restricted to the flat spacetime known as Minkowski space. A locally Lorentz-invariant frame that abides by special relativity can be defined at sufficiently small scales, even in curved spacetime.

Galileo Galilei had already postulated that there is no absolute and well-defined state of rest (no privileged reference frames), a principle now called Galileo's principle of relativity. Einstein extended this principle so that it accounted for the constant speed of light, a phenomenon that had been recently observed in the Michelson–Morley experiment. He also postulated that it holds for all the laws of physics, including both the laws of mechanics and of electrodynamics.

Albert Einstein around 1905, the year his "*Annus Mirabilis* papers" – which included *Zur Elektrodynamik bewegter Körper*, the paper founding special relativity – were published.

Postulates

> "Reflections of this type made it clear to me as long ago as shortly after 1900, i.e., shortly after Planck's trailblazing work, that neither mechanics nor electrodynamics could (except in limiting cases) claim exact validity. Gradually I despaired of the possibility of discovering the true laws by means of constructive efforts based on known facts. The longer and the more desperately I tried, the more I came to the conviction that only the discovery of a universal formal principle could lead us to assured results... How, then, could such a universal principle be found?"
>
> — *Albert Einstein: Autobiographical Notes*

Einstein discerned two fundamental propositions that seemed to be the most assured, regardless of the exact validity of the (then) known laws of either mechanics or electrodynamics. These propositions were the constancy of the speed of light and the independence of physical laws (especially the constancy of the speed of light) from the choice of inertial system. In his initial presentation of special relativity in 1905 he expressed these postulates as:

- The Principle of Relativity – The laws by which the states of physical systems undergo change are not affected, whether these changes of state be referred to the one or the other of two systems in uniform translatory motion relative to each other.
- The Principle of Invariant Light Speed – "... light is always propagated in empty space with a definite velocity [speed] *c* which is independent of the state of motion of the emitting body" (from the preface). That is, light in vacuum propagates with the speed *c* (a fixed constant, independent of direction) in at least one system of inertial coordinates (the "stationary system"), regardless of the state of motion of the light source.

The derivation of special relativity depends not only on these two explicit postulates, but also on several tacit assumptions (made in almost all theories of physics), including the isotropy and homogeneity of space and the independence of measuring rods and clocks from their past history.

Following Einstein's original presentation of special relativity in 1905, many different sets of postulates have been proposed in various alternative derivations. However, the most common set of postulates remains those employed by Einstein in his original paper. A more mathematical statement of the Principle of Relativity made later by Einstein, which introduces the concept of simplicity not mentioned above is:

Special principle of relativity: If a system of coordinates K is chosen so that, in relation to it, physical laws hold good in their simplest form, the *same* laws hold good in relation to any other system of coordinates K' moving in uniform translation relatively to K.

Henri Poincaré provided the mathematical framework for relativity theory by proving that Lorentz transformations are a subset of his Poincaré group of symmetry transformations. Einstein later derived these transformations from his axioms.

Many of Einstein's papers present derivations of the Lorentz transformation based upon these two principles.

Einstein consistently based the derivation of Lorentz invariance (the essential core of special relativity) on just the two basic principles of relativity and light-speed invariance. He wrote:

> "The insight fundamental for the special theory of relativity is this: The assumptions relativity and light speed invariance are compatible if relations of a new type ("Lorentz transformation") are postulated for the conversion of coordinates and times of events... The universal principle of the special theory of relativity is contained in the postulate: The laws of physics are invariant with respect to Lorentz transformations (for the transition from one inertial system to any other arbitrarily chosen inertial system). This is a restricting principle for natural laws..."

Thus many modern treatments of special relativity base it on the single postulate of universal Lorentz covariance, or, equivalently, on the single postulate of Minkowski spacetime.

From the principle of relativity alone without assuming the constancy of the speed of light (i.e. using the isotropy of space and the symmetry implied by the principle of special relativity) one can show that the spacetime transformations between inertial frames are either Euclidean, Galilean, or Lorentzian. In the Lorentzian case, one can then obtain relativistic interval conservation and a certain finite limiting speed. Experiments suggest that this speed is the speed of light in vacuum.

The constancy of the speed of light was motivated by Maxwell's theory of electromagnetism and the lack of evidence for the luminiferous ether. There is conflicting evidence on the extent to which Einstein was influenced by the null result of the Michelson–Morley experiment. In any case, the null result of the Michelson–Morley experiment helped the notion of the constancy of the speed of light gain widespread and rapid acceptance.

Lack of an Absolute Reference Frame

The principle of relativity, which states that there is no preferred inertial reference frame, dates back to Galileo, and was incorporated into Newtonian physics. However, in the late 19th century, the existence of electromagnetic waves led physicists to suggest that the universe was filled with a substance that they called "aether", which would act as the medium through which these waves, or vibrations travelled. The aether was thought to constitute an absolute reference frame against which speeds could be measured, and could be considered fixed and motionless. Aether supposedly possessed some wonderful properties: it was sufficiently elastic to support electromagnetic waves, and those waves could interact with matter, yet it offered no resistance to bodies passing through it. The results of various experiments, including the Michelson–Morley experiment, led to the theory of special relativity, by showing that there was no aether. Einstein's solution was to discard the notion of an aether and the absolute state of rest. In relativity, any reference frame moving with uniform motion will observe the same laws of physics. In particular, the speed of light in vacuum is always measured to be c, even when measured by multiple systems that are moving at different (but constant) velocities.

Reference Frames, Coordinates, and the Lorentz Transformation

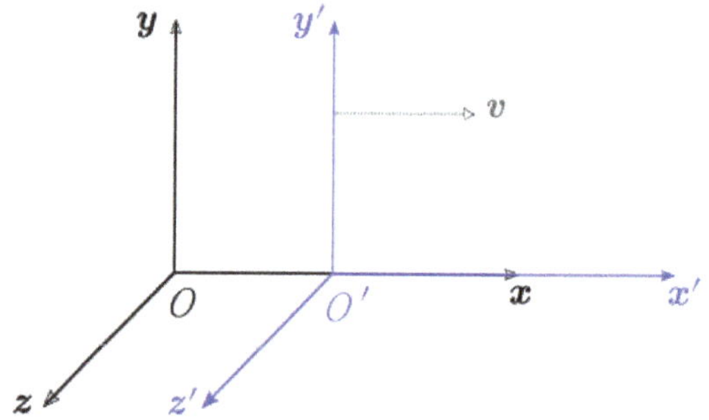

The primed system is in motion relative to the unprimed system with constant velocity v only along the x-axis, from the perspective of an observer stationary in the unprimed system. By the principle of relativity, an observer stationary in the primed system will view a likewise construction except that the velocity they record will be −v. The changing of the speed of propagation of interaction from infinite in non-relativistic mechanics to a finite value will require a modification of the transformation equations mapping events in one frame to another.

Reference frames play a crucial role in relativity theory. The term reference frame as used here is an observational perspective in space which is not undergoing any change in motion (acceleration), from which a position can be measured along 3 spatial axes. In addition, a reference frame has the ability to determine measurements of the time of events using a 'clock' (any reference device with uniform periodicity).

An event is an occurrence that can be assigned a single unique time and location in space relative to a reference frame: it is a "point" in spacetime. Since the speed of light is constant in relativity in each and every reference frame, pulses of light can be used to unambiguously measure distances and refer back the times that events occurred to the clock, even though light takes time to reach the clock after the event has transpired.

For example, the explosion of a firecracker may be considered to be an "event". We can completely specify an event by its four spacetime coordinates: The time of occurrence and its 3-dimensional spatial location define a reference point. Let's call this reference frame *S*.

In relativity theory we often want to calculate the position of a point from a different reference point.

Suppose we have a second reference frame S', whose spatial axes and clock exactly coincide with that of S at time zero, but it is moving at a constant velocity v with respect to S along the x-axis.

Since there is no absolute reference frame in relativity theory, a concept of 'moving' doesn't strictly exist, as everything is always moving with respect to some other reference frame. Instead, any two frames that move at the same speed in the same direction are said to be *comoving*. Therefore, S and S' are not *comoving*.

Define the event to have spacetime coordinates (t,x,y,z) in system S and (t',x',y',z') in S'. Then the Lorentz transformation specifies that these coordinates are related in the following way:

$$\begin{aligned} t' &= \gamma\,(t - vx/c^2) \\ x' &= \gamma\,(x - vt) \\ y' &= y \\ z' &= z, \end{aligned}$$

where

$$\gamma = \frac{1}{\sqrt{1 - \frac{v^2}{c^2}}}$$

is the Lorentz factor and c is the speed of light in vacuum, and the velocity v of S' is parallel to the x-axis. The y and z coordinates are unaffected; only the x and t coordinates are transformed. These Lorentz transformations form a one-parameter group of linear mappings, that parameter being called rapidity.

There is nothing special about the x-axis, the transformation can apply to the y or z axes, or indeed in any direction, which can be done by directions parallel to the motion (which are warped by the γ factor) and perpendicular.

A quantity invariant under Lorentz transformations is known as a Lorentz scalar.

Writing the Lorentz transformation and its inverse in terms of coordinate differences, where for instance one event has coordinates (x_1, t_1) and (x'_1, t'_1), another event has coordinates (x_2, t_2) and (x'_2, t'_2), and the differences are defined as

$$\begin{aligned} \Delta x' &= x'_2 - x'_1\,, \quad & \Delta x &= x_2 - x_1\,, \\ \Delta t' &= t'_2 - t'_1\,, \quad & \Delta t &= t_2 - t_1\,, \end{aligned}$$

we get

$$\Delta x' = \gamma\,(\Delta x - v\Delta t)\,, \quad \Delta x = \gamma\,(\Delta x' + v\Delta t')\,,$$
$$\Delta t' = \gamma\left(\Delta t - \frac{v\Delta x}{c^2}\right), \quad \Delta t = \gamma\left(\Delta t' + \frac{v\Delta x'}{c^2}\right).$$

These effects are not merely appearances; they are explicitly related to our way of measuring *time intervals* between events which occur at the same place in a given coordinate system (called "co-local" events). These time intervals will be *different* in another coordinate system moving with respect to the first, unless the events are also simultaneous. Similarly, these effects also relate to our measured distances between separated but simultaneous events in a given coordinate system of choice. If these events are not co-local, but are separated by distance (space), they will *not* occur at the same *spatial distance* from each other when seen from another moving coordinate system. However, the spacetime interval will be the same for all observers.

Consequences Derived from the Lorentz Transformation

The consequences of special relativity can be derived from the Lorentz transformation equations. These transformations, and hence special relativity, lead to different physical predictions than those of Newtonian mechanics when relative velocities become comparable to the speed of light. The speed of light is so much larger than anything humans encounter that some of the effects predicted by relativity are initially counterintuitive.

Relativity of Simultaneity

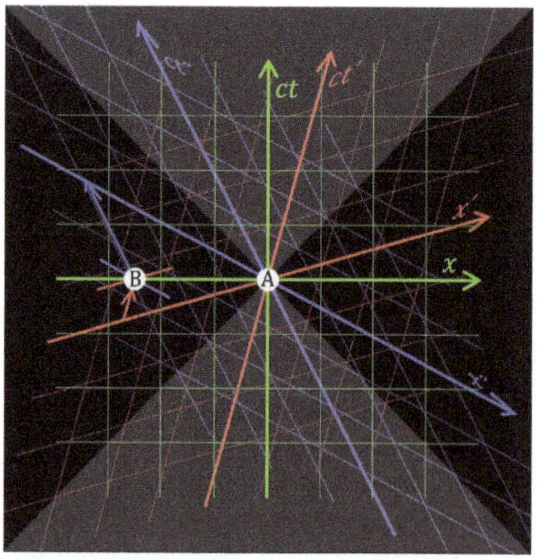

Event B is simultaneous with A in the green reference frame, but it occurs before A in the blue frame, and occurs after A in the red frame.

Two events happening in two different locations that occur simultaneously in the reference frame of one inertial observer, may occur non-simultaneously in the reference frame of another inertial observer (lack of absolute simultaneity).

From the first equation of the Lorentz transformation in terms of coordinate differences

$$\Delta t' = \gamma\left(\Delta t - \frac{v\Delta x}{c^2}\right)$$

it is clear that two events that are simultaneous in frame S (satisfying $\Delta t = 0$), are not necessarily simultaneous in another inertial frame S' (satisfying $\Delta t' = 0$). Only if these events are additionally co-local in frame S (satisfying $\Delta x = 0$), will they be simultaneous in another frame S'.

Time Dilation

The time lapse between two events is not invariant from one observer to another, but is dependent on the relative speeds of the observers' reference frames (e.g., the twin paradox which concerns a twin who flies off in a spaceship traveling near the speed of light and returns to discover that his or her twin sibling has aged much more).

Suppose a clock is at rest in the unprimed system S. The location of the clock on two different ticks is then characterized by $\Delta x = 0$. To find the relation between the times between these ticks as measured in both systems, the first equation can be used to find:

$$\Delta t' = \gamma \Delta t \text{ for events satisfying } \Delta x = 0.$$

This shows that the time ($\Delta t'$) between the two ticks as seen in the frame in which the clock is moving (S'), is *longer* than the time (Δt) between these ticks as measured in the rest frame of the clock (S). Time dilation explains a number of physical phenomena; for example, the lifetime of muons produced by cosmic rays impinging on the Earth's atmosphere is measured to be greater than the lifetimes of muons measured in the laboratory.

Length Contraction

The dimensions (e.g., length) of an object as measured by one observer may be smaller than the results of measurements of the same object made by another observer (e.g., the ladder paradox involves a long ladder traveling near the speed of light and being contained within a smaller garage).

Similarly, suppose a measuring rod is at rest and aligned along the x-axis in the unprimed system S. In this system, the length of this rod is written as Δx. To measure the length of this rod in the system S', in which the rod is moving, the distances x' to the end points of the rod must be measured simultaneously in that system S'. In other words, the measurement is characterized by $\Delta t' = 0$, which can be combined with the fourth equation to find the relation between the lengths Δx and $\Delta x'$:

$$\Delta x' = \frac{\Delta x}{\gamma} \quad \text{for events satisfying } \Delta t' = 0 .$$

This shows that the length ($\Delta x'$) of the rod as measured in the frame in which it is moving (S'), is *shorter* than its length (Δx) in its own rest frame (S).

Composition of Velocities

Velocities (speeds) do not simply add. If the observer in S measures an object moving along the x axis at velocity u, then the observer in the S' system, a frame of reference moving at velocity v in the x direction with respect to S, will measure the object moving with velocity u' where (from the Lorentz transformations above):

$$u' = \frac{dx'}{dt'} = \frac{\gamma\,(dx - vdt)}{\gamma\,(dt - vdx / c^2)} = \frac{(dx/dt) - v}{1 - (v/c^2)(dx/dt)} = \frac{u - v}{1 - uv/c^2} .$$

The other frame S will measure:

$$u = \frac{dx}{dt} = \frac{\gamma\,(dx' + vdt')}{\gamma\,(dt' + vdx' / c^2)} = \frac{(dx'/dt') + v}{1 + (v/c^2)(dx'/dt')} = \frac{u' + v}{1 + u'v/c^2} .$$

Notice that if the object were moving at the speed of light in the S system (i.e. $u = c$), then it would also be moving at the speed of light in the S' system. Also, if both u and v are small with respect to the speed of light, we will recover the intuitive Galilean transformation of velocities

$$u' \approx u - v .$$

The usual example given is that of a train (frame S' above) traveling due east with a velocity v with respect to the tracks (frame S). A child inside the train throws a baseball due east with a velocity u' with respect to the train. In nonrelativistic physics, an observer at rest on the tracks will measure the velocity of the baseball (due east) as $u = u' + v$, while in special relativity this is no longer true; instead the velocity of the baseball (due east) is given by the second equation: $u = (u' + v)/(1 + u'v/c^2)$. Again, there is nothing special about the x or east directions. This formalism applies to any direction by considering parallel and perpendicular components of motion to the direction of relative velocity v.

Other Consequences

Thomas Rotation

The orientation of an object (i.e. the alignment of its axes with the observer's axes) may be different for different observers. Unlike other relativistic effects, this effect becomes quite significant at fairly low velocities as can be seen in the spin of moving particles.

Equivalence of Mass and Energy

As an object's speed approaches the speed of light from an observer's point of view, its relativistic mass increases thereby making it more and more difficult to accelerate it from within the observer's frame of reference.

The energy content of an object at rest with mass m equals mc^2. Conservation of energy implies that, in any reaction, a decrease of the sum of the masses of particles must be accompanied by an increase in kinetic energies of the particles after the reaction. Similarly, the mass of an object can be increased by taking in kinetic energies.

In addition to the papers referenced above—which give derivations of the Lorentz transformation and describe the foundations of special relativity—Einstein also wrote at least four papers giving heuristic arguments for the equivalence (and transmutability) of mass and energy, for $E = mc^2$.

Mass–energy equivalence is a consequence of special relativity. The energy and momentum, which are separate in Newtonian mechanics, form a four-vector in relativity, and this relates the time component (the energy) to the space components (the momentum) in a non-trivial way. For an object at rest, the energy–momentum four-vector is $(E/c, 0, 0, 0)$: it has a time component which is the energy, and three space components which are zero. By changing frames with a Lorentz transformation in the x direction with a small value of the velocity v, the energy momentum four-vector becomes $(E/c, Ev/c^2, 0, 0)$. The momentum is equal to the energy multiplied by the velocity divided by c^2. As such, the Newtonian mass of an object, which is the ratio of the momentum to the velocity for slow velocities, is equal to E/c^2.

The energy and momentum are properties of matter and radiation, and it is impossible to deduce that they form a four-vector just from the two basic postulates of special relativity by themselves, because these don't talk about matter or radiation, they only talk about space and time. The derivation therefore requires some additional physical reasoning. In his 1905 paper, Einstein used the additional principles that Newtonian mechanics should hold for slow velocities, so that there is one energy scalar and one three-vector momentum at slow velocities, and that the conservation law for energy and momentum is exactly true in relativity. Furthermore, he assumed that the energy of light is transformed by the same Doppler-shift factor as its frequency, which he had previously shown to be true based on Maxwell's equations. The first of Einstein's papers on this subject was "Does the Inertia of a Body Depend upon its Energy Content?" in 1905. Although Einstein's argument in this paper is nearly universally accepted by physicists as correct, even self-evident, many authors over the years have suggested that it is wrong. Other authors suggest that the argument was merely inconclusive because it relied on some implicit assumptions.

Einstein acknowledged the controversy over his derivation in his 1907 survey paper on special relativity. There he notes that it is problematic to rely on Maxwell's equations for the heuristic mass–energy argument. The argument in his 1905 paper can be carried out with the emission of any massless particles, but the Maxwell equations are implicitly used to make it obvious that the emission of light in particular can be achieved only by doing work. To emit electromagnetic waves, all you have to do is shake a charged particle, and this is clearly doing work, so that the emission is of energy.

How Far can one Travel from the Earth?

Since one can not travel faster than light, one might conclude that a human can never travel farther from Earth than 40 light years if the traveller is active between the ages of 20 and 60. One would easily think that a traveller would never be able to reach more than the very few solar systems which exist within the limit of 20–40 light years from the earth. But that would be a mistaken conclusion. Because of time dilation, a hypothetical spaceship can travel thousands of light years during the pilot's 40 active years. If a spaceship could be built that accelerates at a constant 1 *g*, it will, after a little less than a year, be travelling at almost the speed of light as seen from Earth. This is described by:

$$v(t) = \frac{at}{\sqrt{1 + \frac{a^2t^2}{c^2}}}$$

where v(t) is the velocity at a time, t, a is the acceleration of 1g and t is the time as measured by people on Earth. Therefore, after 1 year of accelerating at 9.81 m/s^2, the spaceship will be travelling at v = 0.77c relative to Earth. Time dilation will increase the travellers life span as seen from the reference frame of the Earth to 2.7 years, but his lifespan measured by a clock travelling with him will not change. During his journey, people on Earth will experience more time than he does. A 5-year round trip for him will take 6½ Earth years and cover a distance of over 6 light-years. A 20-year round trip for him (5 years accelerating, 5 decelerating, twice each) will land him back on Earth having travelled for 335 Earth years and a distance of 331 light years. A full 40-year trip at 1 *g* will appear on Earth to last 58,000 years and cover a distance of 55,000 light years. A 40-year trip at 1.1 *g* will take 148,000 Earth years and cover about 140,000 light years. A one-way 28 year (14 years accelerating, 14 decelerating as measured with the cosmonaut's clock) trip at 1 *g* acceleration could reach 2,000,000 light-years to the Andromeda Galaxy. This same time dilation is why a muon travelling close to *c* is observed to travel much further than *c* times its half-life (when at rest).

Causality and Prohibition of Motion Faster than Light

In diagram the interval AB is 'time-like'; i.e., there is a frame of reference in which events A and B occur at the same location in space, separated only by occurring at different times. If A precedes B in that frame, then A precedes B in all frames. It is hypothetically possible for matter (or information) to travel from A to B, so there can be a causal relationship (with A the cause and B the effect).

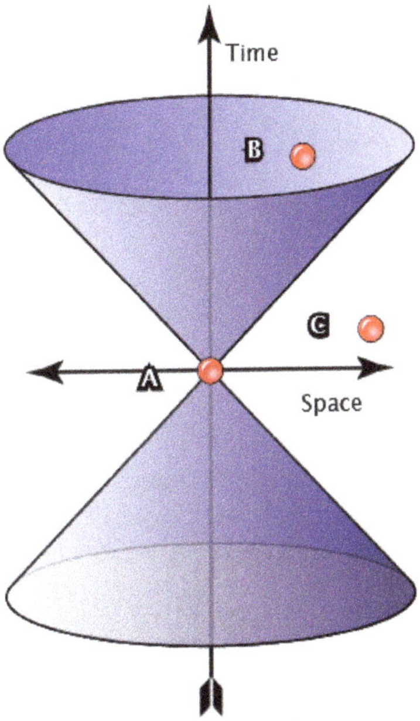

Diagram Light cone

The interval AC in the diagram is 'space-like'; i.e., there is a frame of reference in which events A and C occur simultaneously, separated only in space. There are also frames in which A precedes C (as shown) and frames in which C precedes A. If it were possible for a cause-and-effect relationship to exist between events A and C, then paradoxes of causality would result. For example, if A was the cause, and C the effect, then there would be frames of reference in which the effect preceded the cause. Although this in itself won't give rise to a paradox, one can show that faster than light signals can be sent back into one's own past. A causal paradox can then be constructed by sending the signal if and only if no signal was received previously.

Therefore, if causality is to be preserved, one of the consequences of special relativity is that no information signal or material object can travel faster than light in vacuum. However, some "things" can still move faster than light. For example, the location where the beam of a search light hits the bottom of a cloud can move faster than light when the search light is turned rapidly.

Even without considerations of causality, there are other strong reasons why faster-than-light travel is forbidden by special relativity. For example, if a constant force is applied to an object for a limitless amount of time, then integrating $F = dp/dt$ gives a momentum that grows without bound, but this is simply because $p = m\gamma v$ approaches infinity as v approaches c. To an observer who is not accelerating, it appears as though the object's inertia is increasing, so as to produce a smaller acceleration in response to the same force. This behavior is observed in particle accelerators, where each charged particle is accelerated by the electromagnetic force.

Geometry of Spacetime

Comparison between Flat Euclidean Space and Minkowski Space

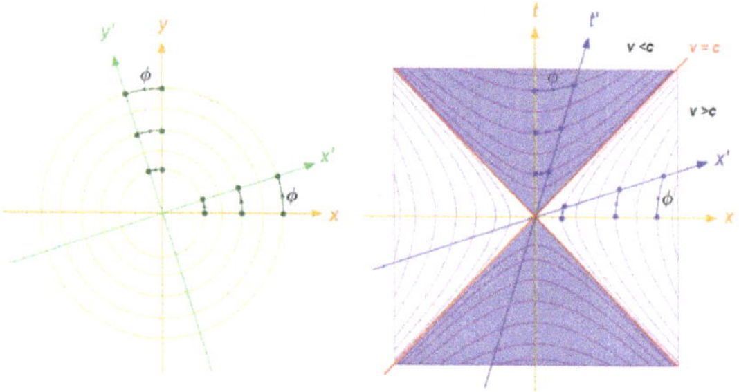

Orthogonality and rotation of coordinate systems compared between left: Euclidean space through circular angle φ, right: in Minkowski spacetime through hyperbolic angle ϕ (red lines labelled *c* denote the worldlines of a light signal, a vector is orthogonal to itself if it lies on this line).

Special relativity uses a 'flat' 4-dimensional Minkowski space – an example of a spacetime. Minkowski spacetime appears to be very similar to the standard 3-dimensional Euclidean space, but there is a crucial difference with respect to time.

In 3D space, the differential of distance (line element) *ds* is defined by

$$ds^2 = d\mathbf{x} \cdot d\mathbf{x} = dx_1^2 + dx_2^2 + dx_3^2,$$

where $dx = (dx_1, dx_2, dx_3)$ are the differentials of the three spatial dimensions. In Minkowski geometry, there is an extra dimension with coordinate X^0 derived from time, such that the distance differential fulfills

$$ds^2 = -dX_0^2 + dX_1^2 + dX_2^2 + dX_3^2,$$

where $dX = (dX_0, dX_1, dX_2, dX_3)$ are the differentials of the four spacetime dimensions. This suggests a deep theoretical insight: special relativity is simply a rotational symmetry of our spacetime, analogous to the rotational symmetry of Euclidean space. Just as Euclidean space uses a Euclidean metric, so spacetime uses a Minkowski metric. Basically, special relativity can be stated as the *invariance of any spacetime interval* (that is the 4D distance between any two events) when viewed from *any inertial reference frame*. All equations and effects of special relativity can be derived from this rotational symmetry (the Poincaré group) of Minkowski spacetime.

The actual form of *ds* above depends on the metric and on the choices for the X^0 coordinate. To make the time coordinate look like the space coordinates, it can be treated as imaginary: $X_0 = ict$ (this is called a Wick rotation). According to Misner, Thorne and Wheeler (1971, §2.3), ultimately the deeper understanding of both special and general relativity will come from the study of the Minkowski metric (described below) and to take $X^0 = ct$, rather than a "disguised" Euclidean metric using *ict* as the time coordinate.

Some authors use $X^0 = t$, with factors of c elsewhere to compensate; for instance, spatial coordinates are divided by c or factors of $c^{\pm 2}$ are included in the metric tensor. These numerous conventions can be superseded by using natural units where $c = 1$. Then space and time have equivalent units, and no factors of c appear anywhere.

3D Spacetime

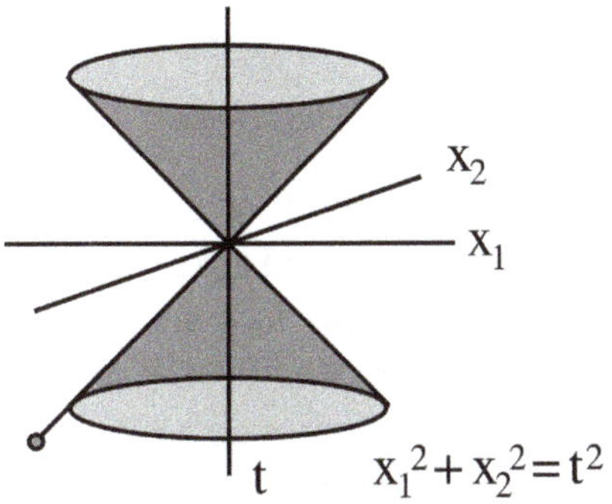

Three-dimensional dual-cone.

If we reduce the spatial dimensions to 2, so that we can represent the physics in a 3D space

$$ds^2 = dx_1^2 + dx_2^2 - c^2dt^2,$$

we see that the null geodesics lie along a dual-cone defined by the equation;

$$ds^2 = 0 = dx_1^2 + dx_2^2 - c^2dt^2$$

or simply

$$dx_1^2 + dx_2^2 = c^2dt^2,$$

which is the equation of a circle of radius $c\,dt$.

4D Spacetime

If we extend this to three spatial dimensions, the null geodesics are the 4-dimensional cone:

$$ds^2 = 0 = dx_1^2 + dx_2^2 + dx_3^2 - c^2dt^2$$

so

$$dx_1^2 + dx_2^2 + dx_3^2 = c^2dt^2.$$

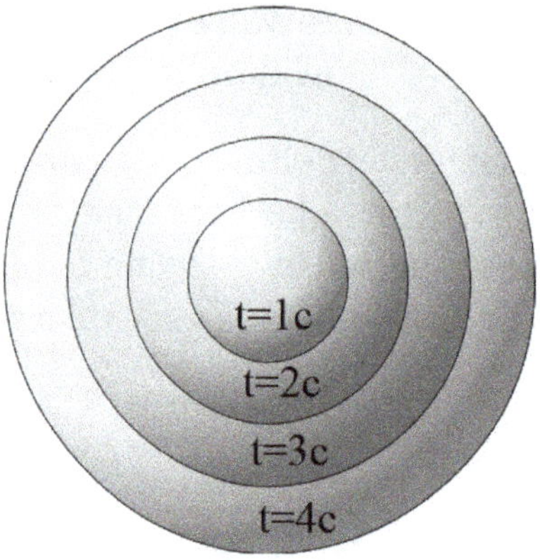

Null spherical space.

This null dual-cone represents the "line of sight" of a point in space. That is, when we look at the stars and say "The light from that star which I am receiving is X years old", we are looking down this line of sight: a null geodesic. We are looking at an event a distance $d = \sqrt{x_1^2 + x_2^2 + x_3^2}$ away and a time d/c in the past. For this reason the null dual cone is also known as the 'light cone'. (The point in the lower left of the picture above right represents the star, the origin represents the observer, and the line represents the null geodesic "line of sight".)

The cone in the $-t$ region is the information that the point is 'receiving', while the cone in the $+t$ section is the information that the point is 'sending'.

The geometry of Minkowski space can be depicted using Minkowski diagrams, which are useful also in understanding many of the thought-experiments in special relativity.

Note that, in 4d spacetime, the concept of the center of mass becomes more complicated.

Physics in Spacetime

Transformations of Physical Quantities between Reference Frames

Above, the Lorentz transformation for the time coordinate and three space coordinates illustrates that they are intertwined. This is true more generally: certain pairs of "timelike" and "spacelike" quantities naturally combine on equal footing under the same Lorentz transformation.

The Lorentz transformation in standard configuration above, i.e. for a boost in the x direction, can be recast into matrix form as follows:

$$\begin{pmatrix} ct' \\ x' \\ y' \\ z' \end{pmatrix} = \begin{pmatrix} \gamma & -\beta\gamma & 0 & 0 \\ -\beta\gamma & \gamma & 0 & 0 \\ 0 & 0 & 1 & 0 \\ 0 & 0 & 0 & 1 \end{pmatrix} \begin{pmatrix} ct \\ x \\ y \\ z \end{pmatrix} = \begin{pmatrix} \gamma ct - \gamma\beta x \\ \gamma x - \beta\gamma ct \\ y \\ z \end{pmatrix}.$$

In Newtonian mechanics, quantities which have magnitude and direction are mathematically described as 3d vectors in Euclidean space, and in general they are parametrized by time. In special relativity, this notion is extended by adding the appropriate timelike quantity to a spacelike vector quantity, and we have 4d vectors, or "four vectors", in Minkowski spacetime. The components of vectors are written using tensor index notation, as this has numerous advantages. The notation makes it clear the equations are manifestly covariant under the Poincaré group, thus bypassing the tedious calculations to check this fact. In constructing such equations, we often find that equations previously thought to be unrelated are, in fact, closely connected being part of the same tensor equation. Recognizing other physical quantities as tensors simplifies their transformation laws. Throughout, upper indices (superscripts) are contravariant indices rather than exponents except when they indicate a square (this is should be clear from the context), and lower indices (subscripts) are covariant indices. For simplicity and consistency with the earlier equations, Cartesian coordinates will be used.

The simplest example of a four-vector is the position of an event in spacetime, which constitutes a timelike component ct and spacelike component x = (x, y, z), in a contravariant position four vector with components:

$$X^{\nu} = (X^0, X^1, X^2, X^3) = (ct, x, y, z) = (ct, \mathbf{x}).$$

where we define $X^0 = ct$ so that the time coordinate has the same dimension of distance as the other spatial dimensions; so that space and time are treated equally. Now the transformation of the contravariant components of the position 4-vector can be compactly written as:

$$X^{\mu'} = \Lambda^{\mu'}{}_{\nu} X^{\nu}$$

where there is an implied summation on ν from 0 to 3, and $\Lambda^{\mu'}{}_{\nu}$ is a matrix.

More generally, all contravariant components of a four-vector T^{ν} transform from one frame to another frame by a Lorentz transformation:

$$T^{\mu'} = \Lambda^{\mu'}{}_{\nu} T^{\nu}$$

Examples of other 4-vectors include the four-velocity U^{μ}, defined as the derivative of the position 4-vector with respect to proper time:

$$U^{\mu} = \frac{dX^{\mu}}{d\tau} = \gamma(v)(c, v_x, v_y, v_z) = \gamma(v)(c, \mathbf{v}).$$

where the Lorentz factor is:

$$\gamma(v) = \frac{1}{\sqrt{1-(v/c)^2}}, \quad v^2 = v_x^2 + v_y^2 + v_z^2.$$

The relativistic energy $E = \gamma(v)mc^2$ and relativistic momentum $\mathbf{p} = \gamma(v)m\mathbf{v}$ of an object are respectively the timelike and spacelike components of a contravariant four momentum vector:

$$P^\mu = mU^\mu = m\gamma(v)(c, v_x, v_y, v_z) = (E/c, p_x, p_y, p_z) = (E/c, \mathbf{p}).$$

where m is the invariant mass.

The four-acceleration is the proper time derivative of 4-velocity:

$$A^\mu = \frac{dU^\mu}{d\tau}.$$

The transformation rules for *three*-dimensional velocities and accelerations are very awkward; even above in standard configuration the velocity equations are quite complicated owing to their non-linearity. On the other hand, the transformation of *four*-velocity and *four*-acceleration are simpler by means of the Lorentz transformation matrix.

The four-gradient of a scalar field φ transforms covariantly rather than contravariantly:

$$\begin{pmatrix} \frac{1}{c}\frac{\partial\phi}{\partial t'} & \frac{\partial\phi}{\partial x'} & \frac{\partial\phi}{\partial y'} & \frac{\partial\phi}{\partial z'} \end{pmatrix} = \begin{pmatrix} \frac{1}{c}\frac{\partial\phi}{\partial t} & \frac{\partial\phi}{\partial x} & \frac{\partial\phi}{\partial y} & \frac{\partial\phi}{\partial z} \end{pmatrix} \begin{pmatrix} \gamma & -\beta\gamma & 0 & 0 \\ -\beta\gamma & \gamma & 0 & 0 \\ 0 & 0 & 1 & 0 \\ 0 & 0 & 0 & 1 \end{pmatrix}.$$

that is:

$$(\partial_{\mu'}\phi) = \Lambda_{\mu'}{}^{\nu}(\partial_\nu\phi), \quad \partial_\mu \equiv \frac{\partial}{\partial x^\mu}.$$

only in Cartesian coordinates. It's the covariant derivative which transforms in manifest covariance, in Cartesian coordinates this happens to reduce to the partial derivatives, but not in other coordinates.

More generally, the covariant components of a 4-vector transform according to the *inverse* Lorentz transformation:

$$\Lambda_{\mu'}{}^{\nu} T^{\mu'} = T^\nu$$

where $\Lambda_{\mu'}{}^{\nu}$ is the reciprocal matrix of $\Lambda^{\mu'}{}_{\nu}$.

The postulates of special relativity constrain the exact form the Lorentz transformation matrices take.

More generally, most physical quantities are best described as (components of) tensors. So to transform from one frame to another, we use the well-known tensor transformation law

$$T^{\alpha'\beta'\cdots\zeta'}_{\theta'\iota'\cdots\kappa'} = \Lambda^{\alpha'}{}_{\mu}\Lambda^{\beta'}{}_{\nu}\cdots\Lambda^{\zeta'}{}_{\rho}\Lambda_{\theta'}{}^{\sigma}\Lambda_{\iota'}{}^{\upsilon}\cdots\Lambda_{\kappa'}{}^{\phi}T^{\mu\nu\cdots\rho}_{\sigma\upsilon\cdots\phi}$$

where $\Lambda_{\chi'}{}^{\psi}$ is the reciprocal matrix of $\Lambda^{\chi'}{}_{\psi}$. All tensors transform by this rule.

An example of a four dimensional second order antisymmetric tensor is the relativistic angular momentum, which has six components: three are the classical angular momentum, and the other three are related to the boost of the center of mass of the system. The derivative of the relativistic angular momentum with respect to proper time is the relativistic torque, also second order antisymmetric tensor.

The electromagnetic field tensor is another second order antisymmetric tensor field, with six components: three for the electric field and another three for the magnetic field. There is also the stress–energy tensor for the electromagnetic field, namely the electromagnetic stress–energy tensor.

Metric

The metric tensor allows one to define the inner product of two vectors, which in turn allows one to assign a magnitude to the vector. Given the four-dimensional nature of spacetime the Minkowski metric η has components (valid in any inertial reference frame) which can be arranged in a 4 × 4 matrix:

$$\eta_{\alpha\beta} = \begin{pmatrix} -1 & 0 & 0 & 0 \\ 0 & 1 & 0 & 0 \\ 0 & 0 & 1 & 0 \\ 0 & 0 & 0 & 1 \end{pmatrix}$$

which is equal to its reciprocal, $\eta^{\alpha\beta}$, in those frames. Throughout we use the signs as above, different authors use different conventions.

The Poincaré group is the most general group of transformations which preserves the Minkowski metric:

$$\eta_{\alpha\beta} = \eta_{\mu'\nu'}\Lambda^{\mu'}{}_{\alpha}\Lambda^{\nu'}{}_{\beta}$$

and this is the physical symmetry underlying special relativity.

The metric can be used for raising and lowering indices on vectors and tensors. Invariants can be constructed using the metric, the inner product of a 4-vector T with another 4-vector S is:

$$T^\alpha S_\alpha = T^\alpha \eta_{\alpha\beta} S^\beta = T_\alpha \eta^{\alpha\beta} S_\beta = \text{invariant scalar}$$

Invariant means that it takes the same value in all inertial frames, because it is a scalar (0 rank tensor), and so no Λ appears in its trivial transformation. The magnitude of the 4-vector T is the positive square root of the inner product with itself:

$$|\mathbf{T}| = \sqrt{T^\alpha T_\alpha}$$

One can extend this idea to tensors of higher order, for a second order tensor we can form the invariants:

$$T^\alpha{}_\alpha, T^\alpha{}_\beta T^\beta{}_\alpha, T^\alpha{}_\beta T^\beta{}_\gamma T^\gamma{}_\alpha = \text{invariant scalars},$$

similarly for higher order tensors. Invariant expressions, particularly inner products of 4-vectors with themselves, provide equations that are useful for calculations, because one doesn't need to perform Lorentz transformations to determine the invariants.

Relativistic Kinematics and Invariance

The coordinate differentials transform also contravariantly:

$$dX^{\mu'} = \Lambda^{\mu'}{}_\nu dX^\nu$$

so the squared length of the differential of the position four-vector dX^μ constructed using

$$d\mathbf{X}^2 = dX^\mu dX_\mu = \eta_{\mu\nu} dX^\mu dX^\nu = -(cdt)^2 + (dx)^2 + (dy)^2 + (dz)^2$$

is an invariant. Notice that when the line element dX^2 is negative that $\sqrt{-d\mathbf{X}^2}$ is the differential of proper time, while when dX^2 is positive, $\sqrt{d\mathbf{X}^2}$ is differential of the proper distance.

The 4-velocity U^μ has an invariant form:

$$\mathbf{U}^2 = \eta_{\nu\mu} U^\nu U^\mu = -c^2,$$

which means all velocity four-vectors have a magnitude of c. This is an expression of the fact that there is no such thing as being at coordinate rest in relativity: at the least, you are always moving forward through time. Differentiating the above equation by τ produces:

$$2\eta_{\mu\nu} A^\mu U^\nu = 0.$$

So in special relativity, the acceleration four-vector and the velocity four-vector are orthogonal.

Relativistic Dynamics and Invariance

The invariant magnitude of the momentum 4-vector generates the energy–momentum relation:

$$\mathbf{P}^2 = \eta^{\mu\nu} P_\mu P_\nu = -(E/c)^2 + p^2.$$

We can work out what this invariant is by first arguing that, since it is a scalar, it doesn't matter in which reference frame we calculate it, and then by transforming to a frame where the total momentum is zero.

$$\mathbf{P}^2 = -(E_{\text{rest}}/c)^2 = -(mc)^2.$$

We see that the rest energy is an independent invariant. A rest energy can be calculated even for particles and systems in motion, by translating to a frame in which momentum is zero.

The rest energy is related to the mass according to the celebrated equation discussed above:

$$E_{\text{rest}} = mc^2.$$

Note that the mass of systems measured in their center of momentum frame (where total momentum is zero) is given by the total energy of the system in this frame. It may not be equal to the sum of individual system masses measured in other frames.

To use Newton's third law of motion, both forces must be defined as the rate of change of momentum with respect to the same time coordinate. That is, it requires the 3D force defined above. Unfortunately, there is no tensor in 4D which contains the components of the 3D force vector among its components.

If a particle is not traveling at c, one can transform the 3D force from the particle's co-moving reference frame into the observer's reference frame. This yields a 4-vector called the four-force. It is the rate of change of the above energy momentum four-vector with respect to proper time. The covariant version of the four-force is:

$$F_\nu = \frac{dP_\nu}{d\tau} = mA_\nu$$

In the rest frame of the object, the time component of the four force is zero unless the "invariant mass" of the object is changing (this requires a non-closed system in which energy/mass is being directly added or removed from the object) in which case it is the negative of that rate of change of mass, times *c*. In general, though, the components of the four force are not equal to the components of the three-force, because the three force is defined by the rate of change of momentum with respect to coordinate time, i.e. dp/dt while the four force is defined by the rate of change of momentum with respect to proper time, i.e. $dp/d\tau$.

In a continuous medium, the 3D *density of force* combines with the *density of power* to form a covariant 4-vector. The spatial part is the result of dividing the force on a small cell (in 3-space) by the volume of that cell. The time component is $-1/c$ times the power transferred to that cell divided by the volume of the cell. This will be used below in the section on electromagnetism.

Relativity and Unifying Electromagnetism

Theoretical investigation in classical electromagnetism led to the discovery of wave propagation. Equations generalizing the electromagnetic effects found that finite propagation speed of the E and B fields required certain behaviors on charged particles. The general study of moving charges forms the Liénard–Wiechert potential, which is a step towards special relativity.

The Lorentz transformation of the electric field of a moving charge into a non-moving observer's reference frame results in the appearance of a mathematical term commonly called the magnetic field. Conversely, the *magnetic* field generated by a moving charge disappears and becomes a purely *electrostatic* field in a comoving frame of reference. Maxwell's equations are thus simply an empirical fit to special relativistic effects in a classical model of the Universe. As electric and magnetic fields are reference frame dependent and thus intertwined, one speaks of *electromagnetic* fields. Special relativity provides the transformation rules for how an electromagnetic field in one inertial frame appears in another inertial frame.

Maxwell's equations in the 3D form are already consistent with the physical content of special relativity, although they are easier to manipulate in a manifestly covariant form, i.e. in the language of tensor calculus.

Status

Special relativity in its Minkowski spacetime is accurate only when the absolute value of the gravitational potential is much less than c^2 in the region of interest. In a strong gravitational field, one must use general relativity. General relativity becomes special relativity at the limit of a weak field. At very small scales, such as at the Planck length and below, quantum effects must be taken into consideration resulting in quantum gravity. However, at macroscopic scales and in the absence of strong gravitational fields, special relativity is experimentally tested to extremely high degree of accuracy (10^{-20}) and thus accepted by the physics community. Experimental results which appear to contradict it are not reproducible and are thus widely believed to be due to experimental errors.

Special relativity is mathematically self-consistent, and it is an organic part of all modern physical theories, most notably quantum field theory, string theory, and general relativity (in the limiting case of negligible gravitational fields).

Newtonian mechanics mathematically follows from special relativity at small velocities (compared to the speed of light) – thus Newtonian mechanics can be considered as a special relativity of slow moving bodies.

Several experiments predating Einstein's 1905 paper are now interpreted as evidence for relativity. Of these it is known Einstein was aware of the Fizeau experiment before 1905, and historians have concluded that Einstein was at least aware of the Michelson–Morley experiment as early as 1899 despite claims he made in his later years that it played no role in his development of the theory.

- The Fizeau experiment (1851, repeated by Michelson and Morley in 1886) measured the speed of light in moving media, with results that are consistent with relativistic addition of colinear velocities.
- The famous Michelson–Morley experiment (1881, 1887) gave further support to the postulate that detecting an absolute reference velocity was not achievable. It should be stated here that, contrary to many alternative claims, it said little about the invariance of the speed of light with respect to the source and observer's velocity, as both source and observer were travelling together at the same velocity at all times.
- The Trouton–Noble experiment (1903) showed that the torque on a capacitor is independent of position and inertial reference frame.
- The Experiments of Rayleigh and Brace (1902, 1904) showed that length contraction doesn't lead to birefringence for a co-moving observer, in accordance with the relativity principle.

Particle accelerators routinely accelerate and measure the properties of particles moving at near the speed of light, where their behavior is completely consistent with relativity theory and inconsistent with the earlier Newtonian mechanics. These machines would simply not work if they were not engineered according to relativistic principles. In addition, a considerable number of modern experiments have been conducted to test special relativity. Some examples:

- Tests of relativistic energy and momentum – testing the limiting speed of particles
- Ives–Stilwell experiment – testing relativistic Doppler effect and time dilation
- Time dilation of moving particles – relativistic effects on a fast-moving particle's half-life

- Kennedy–Thorndike experiment – time dilation in accordance with Lorentz transformations
- Hughes–Drever experiment – testing isotropy of space and mass
- Modern searches for Lorentz violation – various modern tests
- Experiments to test emission theory demonstrated that the speed of light is independent of the speed of the emitter.
- Experiments to test the aether drag hypothesis – no "aether flow obstruction".

Detractors

Despite the success of the Theory of Special Relativity, there are still detractors who insist on the existence of the aether. The basis for this is the experiment performed by Georges Sagnac that produced the Sagnac effect. However, this effect has been proven to reconcile with Special Relativity.

Theories of Relativity and Quantum Mechanics

Special relativity can be combined with quantum mechanics to form relativistic quantum mechanics and Quantum electrodynamics. It is an unsolved problem in physics how *general* relativity and quantum mechanics can be unified, but quantum electrodynamics accounts for special relativity; quantum gravity and a "theory of everything", which require a unification including general relativity too, are active and ongoing areas in theoretical research.

The early Bohr–Sommerfeld atomic model explained the fine structure of alkali metal atoms using both special relativity and the preliminary knowledge on quantum mechanics of the time.

In 1928, Paul Dirac constructed an influential relativistic wave equation, now known as the Dirac equation in his honour, that is fully compatible both with special relativity and with the final version of quantum theory existing after 1926. This equation explained not only the intrinsic angular momentum of the electrons called *spin*, it also led to the prediction of the antiparticle of the electron (the positron), and fine structure could only be fully explained with special relativity. It was the first foundation of *relativistic quantum mechanics*. In non-relativistic quantum mechanics, spin is phenomenological and cannot be explained.

On the other hand, the existence of antiparticles leads to the conclusion that relativistic quantum mechanics is not enough for a more accurate and complete theory of particle interactions. Instead, a theory of particles interpreted as quantized fields, called *quantum field theory*, becomes necessary; in which particles can be created and destroyed throughout space and time.

Defining Momentum in Special Relativity

Desirable Properties of Relativistic Momentum

- In the non-relativistic world, momentum is simply given by $\vec{\mathbf{p}} = m\vec{\mathbf{u}}$, where $\vec{\mathbf{u}}$ is the velocity of the object. This does not work with relativity (as will be seen in the following example), and we have to go for a generalization of this.

- We look for a generalized definition of momentum of the form

$$\vec{\mathbf{p}}(\vec{\mathbf{u}}) = M(u)\vec{\mathbf{u}}, \quad \text{with } \lim_{u \to 0} M(u) = m,$$

where m is the mass of the particle. Note that we have still kept certain desirable properties of momentum:

(i) $\vec{\mathbf{p}}$ is in the same direction as $\vec{\mathbf{u}}$.

(ii) The function $M(u)$) depends only on the magnitude of $\vec{\mathbf{u}}$, and is independent of its direction. This corresponds to isotropy of space.

Elastic Collision of Same Masses and Speeds

Consider the elastic collision of two objects A and B of the same mass m, moving towards each other with speed v. They undergo an elastic collision such that each of them gets deflected through an angle θ , without any change in the speeds. Let us call A and B after collision by C and D, respectively, for convenience.

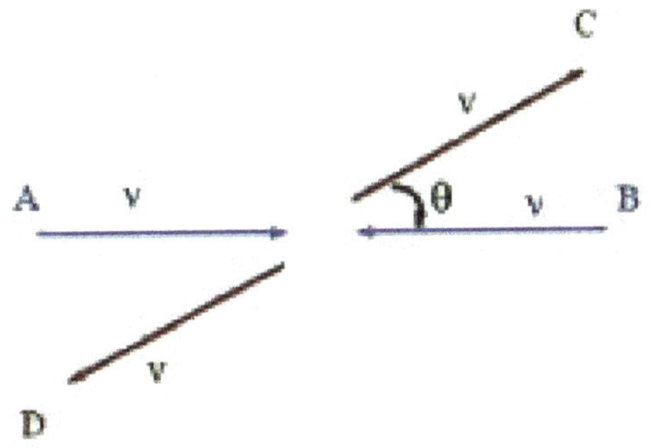

- The momenta of A, B, C, D are then

$$\vec{\mathbf{p}}_A = M(v)(v,0,0), \qquad \vec{\mathbf{p}}_C = M(v)(v\cos\theta, v\sin\theta, 0),$$
$$\vec{\mathbf{p}}_B = M(v)(-v,0,0), \qquad \vec{\mathbf{p}}_D = M(v)(-v\cos\theta, -v\sin\theta, 0)$$

- Clearly, momentum conservation holds here.

Elastic Collision in the Frame of a Stationary Mass

- Let us now see the above collision in the frame S' which is moving with a speed v along x direction. In this frame, B is initially stationary.

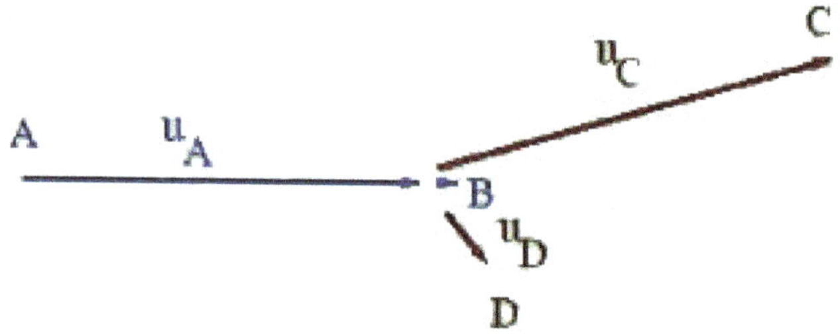

Momenta in the S' Frame

- Since we know how velocities transform under frame change, and since $\vec{\mathbf{p}}$ is a function only of the velocities, we can write down the momenta of the particles in frame S' as

$$\vec{\mathbf{p}}'_A = M(\mathrm{u}_A)\left(\frac{2v}{1+v^2/c^2}, 0, 0\right),$$

$$\vec{\mathbf{p}}'_B = (0,0,0),$$

$$\vec{\mathbf{p}}'_C = M(\mathrm{u}_C)\left(\frac{v+v\cos\theta}{1+v^2\cos\theta/c^2}, \frac{v\sin\theta}{\gamma_v(1+v^2\cos\theta/c^2)}, 0\right),$$

$$\vec{\mathbf{p}}'_D = M(\mathrm{u}_D)\left(\frac{v-v\cos\theta}{1-v^2\cos\theta/c^2}, \frac{-v\sin\theta}{\gamma_v(1-v^2\cos\theta/c^2)}, 0\right)$$

Where $\mathrm{u}_A = 2v/(1+v^2/c^2)$ *and* $\gamma_v = 1/\sqrt{1-v^2/c^2}$

Momentum Conservation Should Hold In S' Too !

- Momentum conservation in y direction gives

$$M(\mathrm{u}_C)\frac{v\sin\theta}{\gamma_v(1+v^2\cos\theta/c^2)} = M(\mathrm{u}_D)\frac{v\sin\theta}{\gamma_v(1-v^2\cos\theta/c^2)}$$

This leads to

$$\frac{M(\mathrm{u}_C)}{M(\mathrm{u}_D)} = \frac{1+v^2\cos\theta/c^2}{1-v^2\cos\theta/c^2}$$

- Now go to the limit of glancing collision, i.e.

$$\theta = 0\,,\mathrm{u}_C = \mathrm{u}_A, \mathrm{u}_D = 0$$

In this limit, the above relation yields

$$\frac{M(\mathrm{u}_A)}{M(0)} = \frac{1+v^2/c^2}{1-v^2/c^2} = \frac{1}{\sqrt{1-\mathrm{u}_A^2/c^2}},$$

the last equality uses $\mathrm{u}_A = 2v/(1+v^2/c^2)$.

- Using $M(0)=m$, we have

$$M(\mathrm{u}_A) = m\Big/\sqrt{1-\mathrm{u}_A^2} = \gamma_{\mathrm{u}_A} m.$$

Relativistic Momentum and Force

- The relativistic momentum is thus

$$\vec{\mathbf{p}} = m\gamma\vec{\mathbf{u}}$$

- This definition is the only one possible which a-llows conservation of momentum to be valid in all frames. Clearly the non-relativistic definition $\vec{\mathbf{p}} = m\vec{\mathbf{u}}$ is not valid at large velocities.

- The relativistic force would naturally be defined as the rate of change of momentum:

$$\vec{\mathbf{F}} = \frac{d\vec{\mathbf{p}}}{dt} = \frac{d}{dt}(m\gamma\vec{\mathbf{u}}).$$

Note that $\vec{\mathbf{F}} \neq m\vec{\mathbf{a}}$, and $\vec{\mathbf{F}} \neq m\gamma\vec{\mathbf{a}}$ either. Indeed in general, force need not even be in the same direction as acceleration.

Defining Relativistic Energy

Kinetic Energy

- Given the momentum $\vec{\mathbf{p}}$, it is possible to define the kinetic energy of a particle as the work done on it while increasing its speed from o to u, in a linear motion.

$$KE = \int \frac{d\vec{\mathbf{p}}}{dt}.d\vec{\mathbf{x}} = \frac{d}{dt}(m\gamma\vec{\mathbf{u}}).\vec{\mathbf{u}}dt$$

$$KE = mc^2(\gamma - 1).$$

- Note that we are still far from saying that the total energy is $m\gamma c^2$ and the rest mass energy is mc^2. We only have an expression for the kinetic energy, and no interpretation.

Energy Conservation in Nuclear Decay

The following argument is a variation on Einstein's original one in his $E = mc^2$ paper.

- Consider a nucleus with mass m_i at rest decaying to another one with mass m_f and two photons leaving in opposite directions with energy E/2 each. The conservation of energy gives

$$E_i = E_f + E$$

where E_f and E_i are the rest energies of the nuclei. Both are at rest by conservation of momentum.

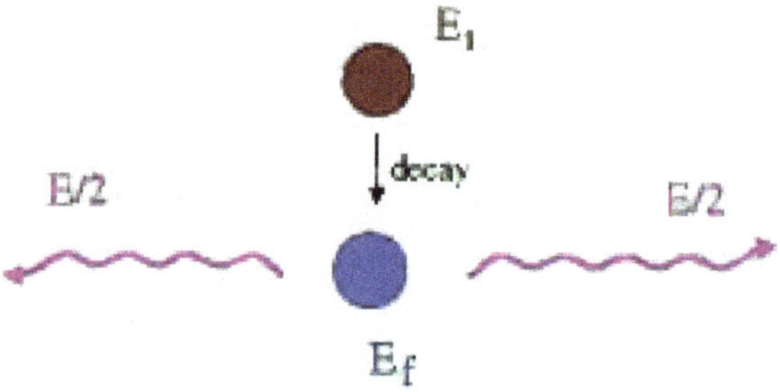

The same Decay in Another Frame

- Now look at the same decay in a frame where the nucleus is moving with speed v and the two photons are emitted along (or opposite) the line of motion.

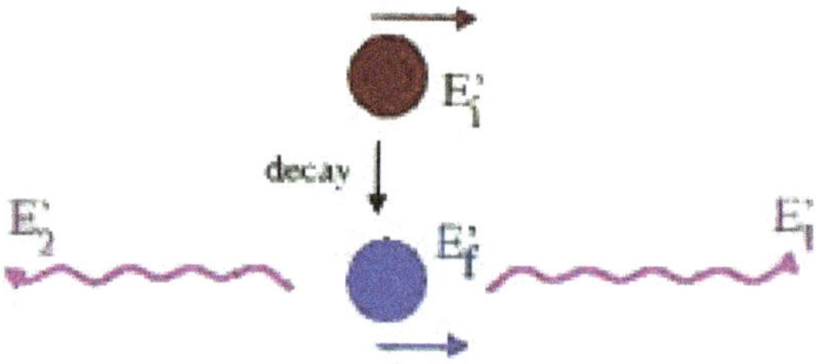

- The energy carried by the photons in this frame is $E_1' = (E/2)\gamma(1-\beta)$, and $E_2' = (E/2)\gamma(1+\beta)$ the total photon energy thus being $E' = E\gamma$. The conservation of energy now gives

$$E_i^{'} = E_f^{'} + \gamma E,$$

where $E_i^{'}$ and $E_f^{'}$ are total energies of the two nuclei.

Postulation of Rest Mass Energy

- The total energy is rest energy plus kinetic energy, i.e.

$$E_i^{'} = E_i + m_i c^2(\gamma - 1), \; E_f^{'} = E_f + m_f c^2(\gamma - 1)$$

- Combining the above equation with energy conservation relations gives

$$m_i c^2 - m_f c^2 = E$$

- This may be "interpreted" as the nuclei having rest energies equal to $m_i c^2$ and $m_f c^2$, respectively, and the difference being emitted as photons.
- This is thus the "postulation" of rest energy of a particle equal to mc^2, which needs to be confirmed by experiments involving radioactive decays: emitted energy and masses of nuclei. The experiments have confirmed this, and hence we have

$$E_{rest} = mc^2$$

Total Relativistic Energy

- The total energy of a relativistic particle is

$$E = E_{rest} + KE = m\gamma c^2.$$

- This can also be written as

$$E = \sqrt{|\vec{\mathbf{p}}|^2 c^2 + m^2 c^4},$$

Tests of Relativistic Energy and Momentum

Tests of relativistic energy and momentum are aimed at measuring the relativistic expressions for energy, momentum, and mass. According to special relativity, the properties of particles moving approximately at the speed of light significantly deviate from the predictions of Newtonian mechanics. For instance, the speed of light cannot be reached by massive particles.

Today, those relativistic expressions for particles close to the speed of light are routinely confirmed in undergraduate laboratories, and necessary in the design and theoretical evaluation of collision experiments in particle accelerators.

Overview

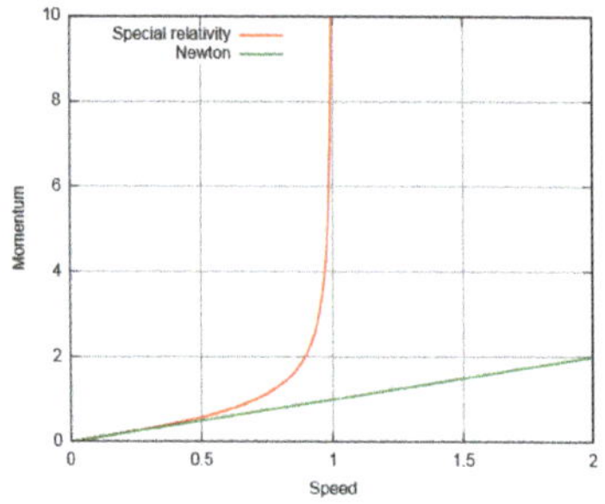

Similar to kinetic energy, relativistic momentum increases to infinity when approaching the speed of light.

In classical mechanics, kinetic energy and momentum are expressed as

$$E_k = \tfrac{1}{2}mv^2, \quad p = mv.$$

On the other hand, special relativity predicts that the speed of light is constant in all inertial frames of references. The relativistic energy–momentum relation reads:

$$E^2 - (pc)^2 = (mc^2)^2,$$

from which the relations for rest energy E_0, relativistic energy (rest + kinetic) E, kinetic energy E_k, and momentum p of massive particles follow:

$$E_0 = mc^2, \quad E = \gamma mc^2, \quad E_k = (\gamma - 1)mc^2, \quad p = \gamma mv,$$

where $\gamma = 1/\sqrt{1-(v/c)^2}$. So relativistic energy and momentum significantly increase with speed, thus the speed of light cannot be reached by massive particles. In some relativity textbooks, the so-called "relativistic mass" $M = \gamma m$ is used as well. However, this concept is considered disadvantageous by many authors, instead the expressions of relativistic energy and momentum should be used to express the velocity dependence in relativity, which provide the same experimental predictions.

Early Experiments

First experiments capable of detecting such relations were conducted by Walter Kaufmann, Alfred Bucherer and others between 1901 and 1915. These experiments were aimed at measuring the deflection of beta rays within a magnetic field so as to determine the mass-to-charge ratio of electrons. Since the charge was known to be velocity independent, any variation had to be attributed to alterations in the electron's momentum or mass (formerly known as transverse electromagnetic mass $m_T = m\gamma$, equivalent to the "relativistic mass" M as indicated above). Since relativistic mass is not often used anymore in modern textbooks, those tests can be described of measurements of relativistic momentum or energy, because the following relation applies:

$$\frac{M}{m} = \frac{p}{mv} = \frac{E}{mc^2} = \gamma$$

Electrons traveling between 0.25–0.75c indicated an increase of momentum in agreement with the relativistic predictions, and were considered as clear confirmations of special relativity. However, it was later pointed out that although the experiments were in agreement with relativity, the precision wasn't sufficient to rule out competing models of the electron, such as the one of Max Abraham.

Already in 1915, however, Arnold Sommerfeld was able to derive the Fine structure of hydrogen-like spectra by using the relativistic expressions for momentum and energy (in the context of the Bohr–Sommerfeld theory). Subsequently, Karl Glitscher simply replaced the relativistic expression's by Abraham's, demonstrating that Abraham's theory is in conflict with experimental data and is therefore refuted, while relativity is in agreement with the data.

Precision Measurements

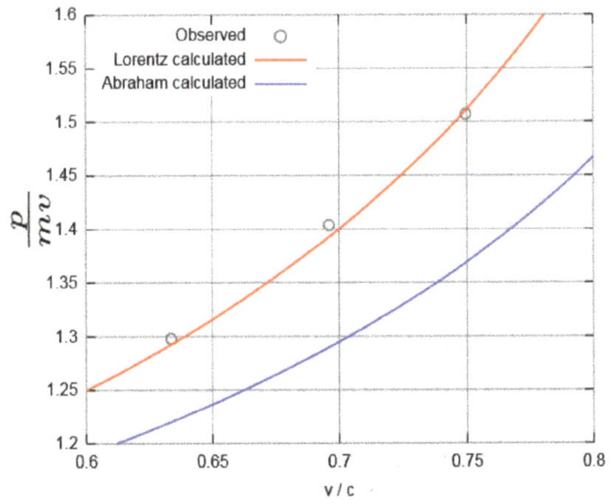

Three data points of Rogers *et al.*, in agreement with special relativity.

In 1940, Rogers *et al.* performed the first electron deflection test sufficiently precise to definitely rule out competing models. As in the Bucherer-Neumann experiments, the velocity and the charge-mass-ratio of beta particles of velocities up to 0.75c was measured. However, they made many improvements, including the employment of a Geiger counter. The accuracy of the experiment by which relativity was confirmed was within 1%.

An even more precise electron deflection test was conducted by Meyer *et al.* (1963). They tested electrons traveling at velocities from 0.987 to 0.99c, which were deflected in a static homogenous magnetic field by which p was measured, and a static cylindrical electric field by which $p^2/(m\gamma)$ was measured. They confirmed relativity with an upper limit for deviations of ~0.00037.

Also measurements of the charge-to-mass ratio and thus momentum of protons have been conducted. Grove and Fox (1953) measured 385-MeV protons moving at ~0.7c. Determination of the angular frequencies and of the magnetic field provided the charge-to-mass ratio. This, together with measuring the magnetic center, allowed to confirm the relativistic expression for the charge-to-mass ratio with a precision of ~0.0006.

However, Zrelov *et al.* (1958) criticized the scant information given by Grove and Fox, emphasizing the difficulty of such measurements due to the complex motion of the protons. Therefore, they conducted a more extensive measurement, in which protons of 660 MeV with mean velocity of 0.8112c were employed. The proton's momentum was measured using a Litz wire, and the velocity was determined by evaluation of Cherenkov radiation. They confirmed relativity with an upper limit for deviations of ~0.0041.

Bertozzi Experiment

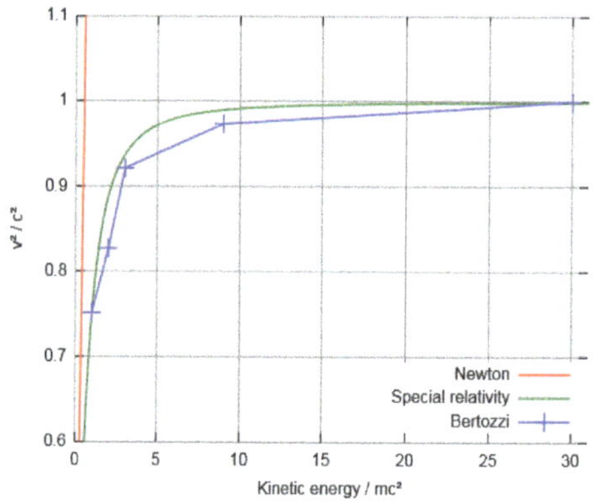

Data of the Bertozzi experiment show close agreement with special relativity. Kinetic energy of five electron runs: 0.5, 1, 1.5, 4.5, 15 MeV (or 1, 2, 3, 9, 30 in mc²). Speed: 0.752, 0.828, 0.922, 0.974, 1.0 in c (or 0.867, 0.910, 0.960, 0.987, 1 in c²).

Since the 1930s, relativity was needed in the construction of particle accelerators, and the precision measurements mentioned above clearly confirmed the theory as well. But those tests demonstrate the relativistic expressions in an indirect way, since many other effects have to be considered in order to evaluate the deflection curve, velocity, and momentum. So an experiment specifically aimed at demonstrating the relativistic effects in a very direct way was conducted by William Bertozzi (1962, 1964).

He employed the electron accelerator facility at MIT in order to initiate five electron runs, with electrons of kinetic energies between 0.5 and 15 MeV. These electrons were produced by a Van de Graaff generator and traveled a distance of 8.4 m, until they hit an aluminium disc. First, the time of flight of the electrons was measured in all five runs – the velocity data obtained were in close agreement with the relativistic expectation. However, at this stage the kinetic energy was only indirectly determined by the accelerating fields. Therefore, the heat produced by some electrons hitting the aluminium disc was measured by calorimetry in order to directly obtain their kinetic energy - those results agreed with the expected energy within 10% error margin.

Undergraduate Experiments

Various experiments have been performed which, due to their simplicity, are still used as undergraduate experiments. Mass, velocity, momentum, and energy of electrons have been measured in different ways in those experiments, all of them confirming relativity. They include experiments involving beta particles, Compton scattering in which electrons exhibit highly relativistic properties and Positron annihilation.

Beta particles	
Marvel *et al.*	2011
Lund *et al.*	2009
Luetzelschwab	2003
Couch *et al.*	1982
Geller *et al.*	1972
Parker	1972
Bartlett *et al.*	1965

Compton recoil electrons	
Jolivette *et al.*	1994
Hoffman	1989
Egelstaff *et al.*	1981
Higbie	1974

Positron annihilation	
Dryzek *et al.*	2006

Particle Accelerators

In modern particle accelerators at high energies, the predictions of special relativity are routinely confirmed, and are necessary for the design and theoretical evaluation of collision experiments, especially in the ultrarelativistic limit. For instance, time dilation of moving particles is necessary to understand the dynamics of particle decay, and the relativistic velocity addition theorem explains the distribution of synchrotron radiation. Regarding the relativistic energy-momentum relations, a series of high precision velocity and energy-momentum experiments have been conducted, in which the energies employed were necessarily much higher than the experiments mentioned above.

Velocity

Time of flight measurements have been conducted to measure differences in the velocities of electrons and light at the SLAC National Accelerator Laboratory. For instance, Brown *et al.* (1973) found no difference in the time of flight of 11-GeV electrons and visible light, setting an upper limit of velocity differences of $\Delta v / c = (-1.3 \pm 2.7) \times 10^{-6}$. Another SLAC experiment conducted by Guiragossián *et al.* (1974) accelerated electrons up to energies of 15 to 20.5 GeV. They used a radio frequency separator (RFS) to measure time-of-flight differences and thus velocity differences between those electrons and 15-GeV gamma rays on a path length of 1015 m. They found no difference, increasing the upper limit to $\Delta v / c = 2\times 10^{-7}$.

Already before, Alväger *et al.* (1964) at the CERN Proton Synchrotron executed a time of flight measurement to test the Newtonian momentum relations for light, being valid in the so-called emission theory. In this experiment, gamma rays were produced in the decay of 6-GeV pions traveling at 0.99975c. If Newtonian momentum $p = mv$ were valid, those gamma rays should have traveled at superluminal speeds. However, they found no difference and gave an upper limit of $\Delta v / c = 10^{-5}$.

Energy and Calorimetry

The intrusion of particles into particle detectors is connected with electron–positron annihilation, Compton scattering, Cherenkov radiation etc., so that a cascade of effects is leading to the production of new particles (photons, electrons, neutrinos, etc.). The energy of such particle showers corresponds to the relativistic kinetic energy and rest energy of the initial particles. This energy can be measured by calorimeters in an electrical, optical, thermal, or acoustical way.

Thermal measurements in order to estimate the relativistic kinetic energy were already carried out by Bertozzi as mentioned above. Additional measurements at SLAC followed, in which the heat produced by 20-GeV electrons was measured in 1982. A beam dump of water-cooled aluminium was employed as calorimeter. The results were in agreement with special relativity, even though the accuracy was only 30%. However, the experimentalists alluded to the fact, that calorimetric tests with 10-GeV electrons were executed already in 1969. There, copper was used as beam dump, and an accuracy of 1% was achieved.

In modern calorimeters called electromagnetic or hadronic depending on the interaction, the energy of the particle showers is often measured by the ionization caused by them. Also excitations can arise in scintillators, whereby light is emitted and then measured by a scintillation counter. Cherenkov radiation is measured as well. In all of those methods, the measured energy is proportional to the initial particle energy.

Annihilation and Pair Production

Relativistic energy and momentum can also be measured by studying processes such as annihilation and pair production. For instance, the rest energy of electrons and positrons is 0.51 MeV respectively. When a photon interacts with an atomic nucleus, electron-positron pairs can be generated in case the energy of the photon matches the required threshold energy, which is the combined electron-positron rest energy of 1.02 MeV. However, if the photon energy is even higher, than the exceeding energy is converted into kinetic energy of the particles. The reverse process occurs in electron-positron annihilation at low energies, in which process photons are created having the same energy as the electron-positron pair. These are direct examples of $E_0 = mc^2$ (mass–energy equivalence).

There are also many examples of conversion of relativistic kinetic energy into rest energy. In 1974, SLAC National Accelerator Laboratory accelerated electrons and positrons up to relativistic velocities, so that their relativistic energy γmc^2 (i.e. the sum of their rest energy and kinetic energy) is significantly increased to about 1500 MeV each. When those particles collide, other particles such as the J/ψ meson of rest energy of about 3000 MeV were produced. Much higher energies were employed at the Large Electron–Positron Collider in 1989, where electrons and positrons were accelerated up to 45 GeV each, in order to produce W and Z bosons of rest energies between 80 and 91 GeV. Later, the energies were considerably increased to 200 GeV to generate pairs of W bosons. Such bosons were also measured using proton-antiproton annihilation. The combined rest energy of those particles amounts to approximately 0.938 GeV each. The Super Proton Synchrotron accelerated those particle up to relativistic velocities and energies of approximately 270 GeV each, so that the center of mass energy at the collision reaches 540 GeV. Thereby, quarks and antiquarks gained the necessary energy and momentum to annihilate into W and Z bosons.

Many other experiments involving the creation of a considerable amount of different particles at relativistic velocities have been (and still are) conducted in hadron colliders such as Tevatron (up to 1 TeV), the Relativistic Heavy Ion Collider (up to 200 GeV), and most recently the Large Hadron Collider (up to 7 TeV) in the course of searching for the Higgs boson.

References

- Edwin F. Taylor & John Archibald Wheeler (1992). Spacetime Physics: Introduction to Special Relativity. W. H. Freeman. ISBN 0-7167-2327-1
- Tom Roberts & Siegmar Schleif (October 2007). "What is the experimental basis of Special Relativity?". Usenet Physics FAQ. Retrieved 2008-09-17
- Bertozzi, William (1964), "Speed and Kinetic Energy of Relativistic Electrons", American Journal of Physics, 32 (7): 551–555, Bibcode:1964AmJPh..32..551B, doi:10.1119/1.1970770
- David Morin (2007) Introduction to Classical Mechanics, Cambridge University Press, Cambridge, chapter 11, Appendix I, ISBN 1-139-46837-5
- Signell, Peter. "Appearances at Relativistic Speeds". Project PHYSNET. Michigan State University, East Lansing, MI. Archived from the original (PDF) on 12 April 2017. Retrieved 12 April 2017
- C.D. Anderson (1933). "The Positive Electron". Phys. Rev. 43 (6): 491–494. Bibcode: 1933 PhRv...43..491A. doi:10.1103/PhysRev.43.491
- Bartlett, A. A.; Correll, Malcolm (1965), "An Undergraduate Laboratory Apparatus for Measuring e/m as a Function of Velocity. I", American Journal of Physics, 33 (4): 327–339, Bibcode:1965AmJPh..33..327B, doi:10.1119/1.1971493

- E. J. Post (1962). Formal Structure of Electromagnetics: General Covariance and Electromagnetics. Dover Publications Inc. ISBN 0-486-65427-3
- Baglio, Julien (26 May 2007). "Acceleration in special relativity: What is the meaning of "uniformly accelerated movement" ?" (PDF). Physics Department, ENS Cachan. Retrieved 22 January 2016
- Meyer, V.; Reichart, W.; Staub, H.H. (1963). "Experimentelle Untersuchung der Massen-Impulsrelation des Elektrons". Helvetica Physica Acta. 36: 981–992. doi:10.5169/seals-113412
- Higbie, J. (1974), "Undergraduate Relativity Experiment", American Journal of Physics, 42 (8): 642–644, Bibcode:1974AmJPh..42..642H, doi:10.1119/1.1987800
- Edwin F. Taylor; John Archibald Wheeler (1992). Spacetime Physics: Introduction to Special Relativity. New York: W. H. Freeman. ISBN 0-7167-2327-1
- Chase, Scott I. "Apparent Superluminal Velocity of Galaxies". The Original Usenet Physics FAQ. Department of Mathematics, University of California, Riverside. Retrieved 12 April 2017
- Alväger, T.; Farley, F. J. M.; Kjellman, J.; Wallin, L. (1964), "Test of the second postulate of special relativity in the GeV region", Physics Letters, 12 (3): 260–262, Bibcode:1964PhL....12..260A, doi:10.1016/0031-9163(64)91095-9

Permissions

All chapters in this book are published with permission under the Creative Commons Attribution Share Alike License or equivalent. Every chapter published in this book has been scrutinized by our experts. Their significance has been extensively debated. The topics covered herein carry significant information for a comprehensive understanding. They may even be implemented as practical applications or may be referred to as a beginning point for further studies.

We would like to thank the editorial team for lending their expertise to make the book truly unique. They have played a crucial role in the development of this book. Without their invaluable contributions this book wouldn't have been possible. They have made vital efforts to compile up to date information on the varied aspects of this subject to make this book a valuable addition to the collection of many professionals and students.

This book was conceptualized with the vision of imparting up-to-date and integrated information in this field. To ensure the same, a matchless editorial board was set up. Every individual on the board went through rigorous rounds of assessment to prove their worth. After which they invested a large part of their time researching and compiling the most relevant data for our readers.

The editorial board has been involved in producing this book since its inception. They have spent rigorous hours researching and exploring the diverse topics which have resulted in the successful publishing of this book. They have passed on their knowledge of decades through this book. To expedite this challenging task, the publisher supported the team at every step. A small team of assistant editors was also appointed to further simplify the editing procedure and attain best results for the readers.

Apart from the editorial board, the designing team has also invested a significant amount of their time in understanding the subject and creating the most relevant covers. They scrutinized every image to scout for the most suitable representation of the subject and create an appropriate cover for the book.

The publishing team has been an ardent support to the editorial, designing and production team. Their endless efforts to recruit the best for this project, has resulted in the accomplishment of this book. They are a veteran in the field of academics and their pool of knowledge is as vast as their experience in printing. Their expertise and guidance has proved useful at every step. Their uncompromising quality standards have made this book an exceptional effort. Their encouragement from time to time has been an inspiration for everyone.

The publisher and the editorial board hope that this book will prove to be a valuable piece of knowledge for students, practitioners and scholars across the globe.

Index

www.ingramcontent.com/pod-product-compliance
Lightning Source LLC
Chambersburg PA
CBHW061953190326
41458CB00009B/2862
9781635497069